# *Keep Listening:*

## *A Patient Perspective on Modern Medicine*

*L V Hannah*

First published as an eBook, October 2015

ISBN: 978-1-326-57799-5

Copyright © L V Hannah 2015

L V Hannah asserts the moral right to be identified as the author of this work

All rights reserved. No part of this publication may be reproduced, stored in a retrieval system, or transmitted, in any form or by any means, electronic, mechanical, photocopying, recording or otherwise, without the prior written permission of the publishers.

*In loving memory of my late father*

*FRCS 1963; DObst RCOG 1959; MB ChB 1957*

"Not only was he modest in material aspiration, [his] unassuming confidence belied his immense ability and skill. A man of high principle and intellect, he was a fierce defender of the disadvantaged and the marginalised. His sense of righteousness made him an ideal representative on the health authority manpower committee. During a session with the hospital chairman, who was seconded from the chemical giant ICI, and who was insistent on reducing the level of nursing staff as a cost-cutting exercise, [he] retorted with his typical sense of humour that trained staff must not be treated as pots of paint.

He epitomised the essence of professional integrity, pragmatism and clinical wisdom of the highest order. These fine qualities were indelibly imprinted in all who worked under him throughout his long, distinguished surgical career."

*Extract taken from The Royal College of Surgeons of England, Plarr's Lives of the Fellows Online.*

# Contents

Prologue

The Verdict

The baby's coming early again

Slow recovery

Loss of my womanhood at 33

No, something is wrong…

Life changing surgery at 38

Another slow recovery

Back to theatre

Feeling dry

My altered normality

Dismissive doctors

Holistic and inclusive medicine

The impact

Hope for the future

# Prologue

Daughters often tend to idolize their fathers and I'm no different.

Growing up in a medical household, with my father being a consultant general and vascular surgeon and my mother an ex ward sister, I harboured early aspirations to follow in my parents footsteps. Or, more specifically, dad's.

I thought that being a surgeon would be most excellent – that was, at least, until dad made it very clear quite early on that there was no use being a woman and being a surgeon unless I wasn't going to get married or have kids!! He made it absolutely clear to us all (mum included) that his patients came first second, third and fourth. So his somewhat outdated and chauvinistic attitude put paid to those early childhood career aspirations.

What I did instead was choose to pursue a career as a lawyer, because it was the only other profession I knew about and which dad thought was OK to be doing as a woman! Little did he realise that I'd be putting in many an *"all-nighter"* during my early career as a single woman in London.

But why write this book?

Well, aside from the fact that dad was an old fashioned male chauvinist, he actually spoke an awful lot of sense. In amongst his late night time ranting and raving at mum about the state of the National Health Service (**NHS**) and the bloody *"interfering"* managers who knew nothing about healthcare, the other thing I always remember him saying was:

*"I tell my students to <u>always</u> listen to the patient ...."*.

As far as he was concerned, as long as the patient could speak coherently, the patient was his best diagnostic tool.

Yet, his common complaint was that he would often see patients who had been misdiagnosed, or left untreated elsewhere, simply because the patient story had not been adequately listened to. So, he listened to the patient, operated wherever necessary, and hey presto the patient was cured.

OK, so it was a bit more complicated than that, with the requisite investigative procedures and examinations where required, but it wasn't rocket science. Far from it – it was straightforward thorough history taking followed by the appropriate action.

Yet how many times do we hear in the press of tragic deaths, of lives that could have been saved, had the doctors listened to the patient, or in the case of children, their parents? Of children sent home with meningitis, being told it's a cold or flu, only to die hours or days later?

Or of adults having been sent away from accident and emergency departments (**A&E**) with stomach cramps with a bit of paracetamol only to present again later with acute appendicitis or peritonitis or, worse, die before being properly treated?

Now, I'm not advocating an investigative laparotomy for every adult presenting at A&E with stomach cramps or immediate hospitalisation for every child with fever and flu like symptoms. But what I am saying, is that somewhere in modern medicine, the skill of listening to the patient, of putting the patient at the heart of healthcare, by making each patient feel like they are important, is all too often lost.

This has nothing to do with a *"cash strapped NHS"* but everything to do with basic common sense, thorough history taking and patient service.

The thing is - people do not, on the whole, present at A&E with acute pain or a sick child because they are drug addicts who want a quick shot of morphine or because they are over anxious mothers' molly coddling their offspring …….. they are usually hard working folk who quite frankly can't afford, and don't want to be there. Yet my personal experience is that had my patient history been taken more carefully, or my presenting symptoms been more carefully listened to, or more fundamentally, my requests in childbirth been taken more seriously, then outcomes would have been better all around. Not only would this have resulted in less time in hospital and fewer repeat admissions, it would also have reduced the financial burden on the healthcare system.

Unfortunately, I have had rather too much experience of both the NHS and private healthcare systems in the UK, as a patient, for my liking. And much to my personal, emotional and financial cost. The time I've missed with my young family and husband and the months I've missed off work due to health problems over the last eight years has been immense.

Valuable family time, and important career time which can never be replaced.

My overriding message in this book is a really simple one.

If **_all_** medics listen to the patient more closely and put the patient back at the heart of modern medicine, making each patient feel like their health is the most important thing in the world (not hitting targets or form filling) then the cost burden on the healthcare system would reduce overnight.

Clearly a large proportion of medics already do this, but my patient experience has been that some (and certainly far too many) don't. And this has cost me, my family, my employers, and the NHS and private insurers dearly.

I have no doubt, from my own patient experiences, that a stronger emphasis on listening within medicine would not only result in more satisfied patients, but would also reduce repeat GP visits and A&E admissions and enable swifter recoveries, which, in turn, would reduce healthcare costs.

What I haven't attempted to examine at all, within this book, are the current funding and other systemic problems facing the NHS.

But I do passionately believe that even a little more time spent listening would result in huge cost savings which, of course, would help enormously.

# The Verdict

I sat there towards the end of my follow up appointment with my (lady) surgeon and my physician, to which I'd travelled for well over three hours to attend. Both were looking at their watches. They were very obviously (but politely) mindful that their next patient was almost due.

But I needed to ask the questions. I needed to ask them, so that I could either pursue my case, or draw a line in the sand and move on……..

*"I think we're done unless you have any more questions for us….."*

my surgeon said, breaking the silence, in a friendly but very matter of fact way.

*"Actually, I do"*

I said.

*"My question is whether or not you feel I've suffered nerve damage?*

*I ask this because I can see that the anorectal physiology test results show that I have a loss of sensation in my rectum….. And this loss of sensation has been confirmed by the tests. In fact you will recall that during the defecating proctogram I wasn't aware of my rectum being filled with the barium paste.*

*And from carrying out some research, I understand that this loss of sensation can be due to pudendal nerve damage. "*

At which point my surgeon responded with:

*"but you would also expect to see loss of anal sphincter tone and your squeezing pressure is fine."*

(Actually on checking my Anorectal physiology results when coming to write this memoir my *"squeezing pressure"* isn't fine because my *"five second squeeze increment"* and my *"involuntary squeeze increment"* are both abnormal and I wasn't able to expel a balloon from my rectum.......But attention to detail is not something that seems to be a strong point of many doctors I've encountered along the way…..)

*"But it doesn't feel fine"*

I continued.

*"And I'm sure this problem stems from around the time of the birth of my second child. In fact I was left for far too long in second stage labour and my baby's head became impacted at the level of the Ischial Spines. And I understand that the Pudendal nerve runs around this area and that it's often damaged in childbirth."*

*"That's right.."*

said my surgeon, who then carried on to say:

*"Foetal impaction is a recognised cause of pudendal neuropathy but it's all a bit of a minefield. Anyone who knows anything about pudendal nerve testing will know that the results are wholly unreliable……"*

*"I see"*

I said, feeling both simultaneously intimidated and confused.

Intimidated because the consultant sat opposite me was fiercely confident and self-assured; and confused because I wondered why on earth anyone ever tested for pudendal neuropathy if it was really that unreliable.

I was, however, very determined to have my questions answered, so I carried on, regardless:

*"But if you could please just put yourself in my shoes for a moment.*

*I'm a forty year old woman, with a six figure earning capacity for the next twenty five years. I probably won't work again full-time because of my low energy levels and frequent need for hospitalisation for IV hydration.*

*I therefore need to know whether I would have ended up this way anyway or whether being left for too long in second stage labour has caused my problems?*

*Because if it has I'm going to take action against the hospital responsible for the delivery..........*

*I know this is almost certainly a controversial topic for you to discuss because the NHS is your employer, to whom you owe a duty, but I feel very aggrieved with what has happened; the loss of opportunity with my career, the financial impact, and the deleterious effects that eight years of recurrent hospitalisations has had on my family.*

*So I'd please like your expert opinion.*

*You see I've had a preliminary expert view which says that on balance, the outcome of my problems would have been the same even without a prolonged second stage of labour. The expert said that the redundant colon, hysterectomy and adhesions from previous surgery were all predisposing factors for the slow transit constipation and that any pudendal neuropathy and prolapse/rectocele caused in childbirth would have had minimal impact. In short his "opinion" was that I would have ended up with a colectomy in any event.*

*But I find that almost impossible to comprehend when I had no problems prior to the birth of my second child....."*

I realised that as I asked this rather convoluted question I sounded incredibly apologetic. And the truth is - I felt it.

On the one hand I was at the consultation asking for help (albeit in my capacity as a private patient) and yet on the other I was threatening to sue the NHS, with whom both consultants were also employed.......

How, therefore, was I realistically expecting my question to be answered? And was there really any point in asking the question at all? (I doubted as much, even before the answer came...)

*"Well it has been argued that the length of the colon has little to do with transit time and prolapse is fairly common in women, and colectomy is more common in women than in men, but admittedly not at your age......"*

said my surgeon.

OK – I thought to myself, but I was still confused, as this didn't really give me any answers so I tried again with a different question:

*"So, what would you do if you were me?"*

(Meaning would you take action if you'd been through the hell that I've been through and feel that you were afforded negligent care in childbirth?) Unfortunately, however, I obviously didn't make my question clear and rather the surgeon had interpreted the question as "*what would you do if you were [instructed as an expert witness by] me*" because the answer that came was this:

*"If I were instructed as an expert witness in your case I would conclude that the outcome would probably be the same regardless of the prolonged second stage of labour. You see I come at it from a different angle altogether really in that there must have been something pretty fundamental wrong with your colon for it to have stopped working in the first place...."*

I thanked the surgeon for the opinion and explained that this would at least enable me to move on, psychologically, from the mind set of taking legal action, but I still felt a whole raft of emotions including anger, sadness and unfairness.

Unfairness that I've been left with a life limiting and life changing disability .... Which, despite the conversation I'd just had in the consulting room, and the draft expert witness report I'd received, I instinctively knew that my health would be in far better shape had I been afforded a better standard of care and had I been listened to along the way. (Not least because I'd been told by the two consultants who have actually carried out my surgeries that the obstructed labours I endured will have almost certainly **_caused_**, or at the very least, **_been a catalyst to_** my health problems.)

In terms of my legal case, it was established that there had been a breach of the duty of care (i.e. negligence) during the birth of my second daughter because I had been left too long in second stage labour for a woman attempting a vaginal delivery after a caesarean section (**VBAC**), but I had yet to establish causation. And without an expert witness report linking my problems directly to the birth of my second daughter there was absolutely no point in trying to pursue a case.

I'd had the head of clinical negligence at one of the leading law firms in the field instructed in my case for the last eighteen months (which was how long it'd taken to get my medical records, go through them, build a chronology and instruct two expert witnesses).

And before I threw in the towel, I had to decide whether to spend the money putting the relevant expert witness through his paces by being questioned by a QC.

I knew that I was more than capable of putting the expert witness through his paces myself (together with my retained solicitor), but as I'm not a clinical negligence lawyer, nor an advocate I wanted to ensure that I hadn't missed anything and for this it was worth paying to get a QC involved.

As consulting a QC wasn't going to be a cheap endeavour (I'd have to pay for him to read the papers, get up to speed and then question the expert in the consultation in addition to paying my solicitor) it was a decision I had to make with my ever patient and long suffering husband. Thankfully he recognised that before I could have absolute closure on any prospective legal case I needed to go through this exercise. My husband recognised that if I didn't do this, then despite what I'd been told in my last consultation, I would always have a niggling doubt of *"what if?"*…….

So there we were: my husband and I sat around a large conference table, in consultation with a leading QC in his field, with one of the leading solicitors in his field, all ready to question this very eminent surgeon who had provided the draft expert witness report.

It was quite clear to me straight away that my QC was quality in terms of asking the right questions and keeping the expert on his toes, but it was equally clear that the expert knew that his duty was to the Court and that he had to work on the basis of a balance of probability ***and not*** on the art of the possible. The hurdle is set very high when it comes to clinical negligence cases …. and rightly so……..

Anyway, without reliving each round of examination and cross examination, what the consultation managed to illicit, was that had the peculiar set of circumstances in terms of the handling of the birth of my second child and my subsequent medical problems occurred ***after*** the birth of a first child instead (and there were no subsequent births) then I would have had a case to answer and causation would have been made out.

***BUT*** because I also had an obstructed delivery for baby number one and the pudendal nerve testing (which was crucial to proving damage) was not carried out both before and after baby number two it was not possible to categorically link my medical problems to the breach of the duty of care at the second birth. As such the expert could not provide me with an expert opinion that would enable me to launch a successful case.

What I really didn't grasp was why ***my evidence*** would hold so little weight? The pudendal nerve damage that I sustained was, without question in my mind, a result of the trauma I suffered during the delivery of my second baby. One of the symptoms I could categorically pinpoint after that delivery was a loss of sensation in my clitoris. (It was not a complete loss of sensation thankfully, but enough of a loss to make climaxing far more difficult than before the birth. Up until the birth of my second child I had enjoyed incredibly powerful orgasms, often two or three times during intimacy with my husband and on a very regular basis, yet after this birth my ability to climax was almost completely frustrated. Likewise, prior to my second child I became very wet with minimal arousal. Yet since the birth of my second daughter I have never become as wet and very rarely enjoy the complete and overwhelming release of an orgasm.)

Further, the loss of sensation in my rectum, which made knowing when I needed to empty my bowels more tricky also occurred after that birth. Additionally, it is well documented that pudendal nerve damage causes constipation, which ultimately resulted in my colectomy – an operation which actually left me close to death due to post-operative complications. ***BUT*** because the anorectal physiology testing I had wasn't carried out after the first labour and before the second labour there was no "*medical*" evidence to prove that this was the case. (It was only carried out many years after my second labour.)

The expert also postulated that whilst he wasn't a genitourinary expert he felt that perhaps the loss of sensation in my Clitoris and difficulty in climaxing could be multifactorial ………and he suggested that I might want to get an opinion from a genitourinary expert.

My QC was, however, quick to comment that to bring an action based solely on this lack of sensation and difficulty in climaxing (having decided that the bowel angle was too difficult to prove) might be disproportionate. (Which I hasten to add I found find quite extraordinary coming from a man. I wonder if it was a man that was suffering with erectile dysfunction or difficulty with ejaculation, he would have felt it disproportionate?!!) Clearly, however, and for reasons, even as a lawyer, I struggle to understand, my evidence, as the primary witness would not have been enough to get me home and dry in a clinical negligence case.

And whilst I considered obtaining expert evidence from a genitourinary consultant to corroborate the sexual dysfunction element of the case, my instinct was that this too would be an uphill struggle and as such I decided that there was little point in pursuing the litigation any further. I therefore informed my lawyers that I would not be seeking such an opinion and thanked them for their time.

In a funny way I was relieved when I'd made this decision, because despite having spent time, money and energy in reaching this conclusion and despite being 100% certain in my own mind that the injuries I sustained during my second labour and the way the birth was handled led to the demise in my health (including my subsequent hysterectomy, colectomy, sexual dysfunction and inability to maintain euvolemia) I didn't actually relish spending the next eighteen, twenty four or however many months embroiled in litigation which would line my lawyers' pockets and cost the NHS (or their insurers) dearly.

Granted, if I succeeded I'd have got a very large pay-out (possibly even seven figures given my relatively young age, qualifications, experience and earning capacity), but that wouldn't have righted any wrongs or given me my health back.

I'd far rather spend the time I would have spent on my witness statement, correspondence and other aspects of the legal process in a positive way by telling my story here in the hope that it will make a difference to someone, somewhere. Whether that's a difference to a medic who begins to listen more; or to a patient who gains strength in the knowledge that they really are ill and do know their body best; or to those who receive treatment from médicins sans frontières who are benefitting financially with each copy of this book that is sold.

So, exactly where does the story of the demise of my health begin?

In the latter stages of my pregnancy with baby number two………..

# The baby's coming early again!

I was almost thirty five weeks into my pregnancy and baby number two wasn't fully cooked for at least another five weeks - yet sat at that desk in my law firm office, what I was definitely feeling were contractions. No, not braxton hicks….. The pain was too much to be braxton hicks…..

*"Ow! Ow! Ow!"* I'm saying out loud at my desk (whilst underneath my breath I'm effing and blinding like a trouper). *"Now I know I've been damn uncomfortable and wishing the weeks away but please don't come yet little girl".* Clearly I was hoping that my baby could hear me and that hearing me she'd somehow stop the contractions!

Yet having given birth to a healthy baby girl, by emergency caesarean at thirty seven weeks three days, weighing in at 7llbs 1oz first time around, I was not altogether surprised when I started feeling these contractions at thirty four and a bit weeks with baby number two.

Obviously I was alarmed - as a baby isn't considered *"full term"* until thirty seven weeks and I didn't want to labour early because my planned caesarean was not booked in for another three and half weeks. BUT I knew my body. What I was feeling were definitely contractions …. so I duly telephoned my chosen birthing hospital and was put on to the midwife in charge. I was advised that I should come straight in for a check-up.

So I closed down my computer, put my out of office automatic email and voicemail on and let my secretary know where I was going….. adding quickly that I hoped it was a false alarm and that I would see her shortly.

On arrival at the hospital I was given a bed and my contractions and the baby's heart rate were monitored using a cardiotocography machine or electronic foetal monitor (**EFM**). The EFM belt was strapped around my waist with one ultrasound transducer placed above my baby girl's heart and the other above her fundus. From the information relayed via the ultrasound transducers to the paper strip the midwife was able to glean that whilst the contractions as observed by the EFM were far stronger than braxton hicks they were very irregular and weren't speeding up at all. The attending doctor, however, erred on the side of caution and admitted me.

Despite me not wanting to stay in hospital, the doctors were concerned about me going into labour early and didn't want this to happen at home. I was therefore kept in for monitoring over the next few days and given that the baby was not yet *"cooked"* I was given an infusion of steroids (betamethasone 12mg) to mature my baby's lungs and nifedipine to slow my uterine contractions and to put off full labour.

The nifedipine certainly seemed to work well in that during my stay, whilst my contractions remained frequent the intensity eased off somewhat and so on the third day I was allowed home, but told to return if the contractions became stronger again. My discharge details recorded the reason for my admission as being a *"threatened pre-term labour"*.

During this admission I asked my consultant for my planned caesarean to be moved forward a week – to thirty seven weeks. (I was currently booked for thirty eight weeks and three days.) I asked for the earlier date because my baby was already measuring big and as my first daughter was born healthy at thirty seven weeks I reasoned that this second baby would be ready early too. More than anything else, though, I was worried that I'd labour before my planned caesarean date and be persuaded to push.

Unfortunately, however, I was told flatly, without any discussion at all, that an earlier date was not a possibility. Reluctantly, I accepted this answer, even though I had been at pains to explain to my short, fat, balding, arrogant and rather obnoxious consultant that my particular fear was that I would labour earlier than my planned date, be persuaded to push and then end up in an emergency caesarean situation all over again which I didn't relish.

As my husband is 6ft 4ins and I'm only 5ft 3ins I grow long, big (but not necessarily heavy) babies. My previous labour ended with my baby girl unable to descend through my pelvis and me being taken to the operating theatre effing and blinding in total agony (as I'd been brainwashed by the National Childbirth Trust that if I just used gas and air I'd have a far greater chance of a natural birth). I was in such agony by the time I got to theatre that I couldn't keep still and the anaesthetist could barely get the needle in my spine to administer the anaesthetic. Unsurprisingly, I was adamant that I didn't want a rerun of this scenario, hence my decision for a planned caesarean ......

Anyway, following my discharge after this premature labour scare and for the following three weeks, I remained at home and continued to have painful and exhausting contractions every day - but as they weren't getting any stronger or more consistently regular I heeded the advice of my consultant and didn't go back to hospital. Instead I just kept wishing away the time to my planned delivery date.

I'd already given up work early upon my discharge following my preterm labour scare and put my toddler in full time day care (as I was incapable of looking after her with the pain and exhaustion levels I was suffering) and at my weekly midwife appointments (which I was now having due to my earlier admission and continuing contractions) I kept requesting for my caesarean date to be moved forward.

I knew my body.

I knew I wouldn't make it to my planned date.

But nobody listened.

My date wasn't moved.

Fast forward to the Friday before my planned caesarean date – I'm nearly there...... My caesarean section is planned for the following Monday. Except I go into labour early – just as I knew I would.

I was lying in bed next to my husband when, at around 23:00 I started to feel a strengthening and an increasing in the regularity of the contractions. As the time ticked on my contractions continued to get more painful and so I took myself off to the hospital, leaving my husband at home with my daughter, with the plan of my husband following on once my mother arrived. (I still to this day shudder when I think of myself driving in the pitch black along country roads and in pain to the hospital...)

Then, on arrival at the hospital at around some time after 01:00 I'm told I'm about 3cm dilated and that I could either go for a caesarean as soon as a theatre, anaesthetist and obstetrician become available (as I was down as a planned caesarean) or I could attempt a VBAC. The attending doctor strongly recommended that I opt for the VBAC and reassured me that the likelihood of me ending up with another emergency caesarean was slim. (I'd never met him before and had serious doubts as to whether he'd actually read the details of my first labour.)

My initial reaction to his recommendation was to stick to my guns and say no, but the doctor then reminded me of the longer hospital stay and recovery time and the inability to drive for four to six weeks if I opted for the caesarean straight away – whilst with a VBAC I could be home the next day and driving. The attending doctor also explained that I wouldn't get a consultant to do the caesarean if I opted for it straight away as it was out of hours and that it would be the on call registrar who would do the delivery. (I was angered by this because I had been previously assured by my consultant with whom I undertook all my scans privately that he would do the operation, no matter what).

Anyway – what can I say? All of a sudden in between the pains of full blown labour contractions and after an exhausting three weeks of irregular, but still quite painful *"pre labour"* contractions, a VBAC started to become particularly appealing (despite me being adamant all through my pregnancy that the only birth I would consider was a planned caesarean). Before I agreed, though, I asked for a scan because I was concerned that if the baby was back to back with me I didn't want to try for a natural labour. I was, however, informed that a scan wouldn't be much use and I was reassured that things can change as the labour progresses anyway.......

So in the agonising pain of labour I'm eventually swayed to try a VBAC (cue no listening, nor any thorough review of my history .......and despite my clear request for a caesarean recorded on my file). I clearly had a lapse of reason amongst the pain to agree with the junior doctor that a VBAC was a sensible option.

But sod's law. My baby was back to back and didn't turn in labour....It was *"groundhog day"* all over again.

I progressed to 10cm dilated quite quickly, exactly as I had done in my first labour (but this time I decided that I shouldn't just rely on gas and air) and I had an epidural when the pain levels became unbearable. But as Lady Luck was not shining down on me the epidural failed to work on the right side of my body. Oh joy! BUT more alarmingly not only did my baby decide to remain back to back with me (as I feared she would) she also decided not to descend through my pelvis and became lodged at the level of the ischial spines. And I was left in second stage labour for far too long.

My notes recorded that I was fully dilated at 10:00 but that I was not taken to theatre for the spinal anaesthetic until 13:45. Further, whilst a decision was made at 12:25 to take me to theatre for a section I was not actually anaesthetised and ready for the surgery for a further full hour and twenty minutes. So I was left in absolute agony for all that time…….. and I vividly remember screaming in pain and begging them to hurry up…..

Fairly recently when I read through my notes after they were disclosed for the purpose of litigation I noticed that a manuscript amendment had been made to the time I was in theatre for my spinal to 12:45 from 13:45 but there is absolutely no doubt in my mind that it was 13:45 and not 12:45. For a VBAC you're only meant to push for an hour and I was left for nearly four …..

Anyway as you can imagine the theatre experience was far from pleasant as I was in total agony by the time the anaesthetist was ready to numb me. In fact I could barely keep still for the spinal anaesthetic to be administered and when I was finally numbed and on the table, cut open and ready for the delivery the obstetrician couldn't actually deliver my baby because her head had become impacted.

I'd been left for so long with no real progression in descent that my precious baby's head was stuck. Rather impatiently I heard the obstetrician sternly asking the midwife to push my baby back up into my tummy to enable delivery, but I could hear her saying that she couldn't and that she was stuck.

And so the obstetrician then had to get down the business end and do it himself. (As anyone who's had a spinal anaesthetic for a caesarean knows - whilst you feel no pain you can still be aware of some sensations and to this day I can still conjure up the feeling of my little girl being pushed by her head and pulled by her legs free from my womb.) Then to make things worse when my baby was eventually pulled free she took what felt like forever to cry, scoring only 6 on her apgar.

The whole experience was utterly traumatic and unsurprisingly, after three weeks of contractions, followed by a full labour and emergency section I was totally spent. I then started to shake uncontrollably in the hours following the delivery, which the midwife told me was a side effect of the spinal anaesthesia, but with hindsight was almost certainly shock as I was struggling to maintain consciousness through the shaking. But I did remain conscious and the shaking eventually subsided.

And inevitably the pain, shaking and exhaustion was, of course, worth it …….. I had the perfect gift - my healthy bouncing baby girl, weighing in at 7llbs 3oz at just under 38 weeks.

# Slow recovery

Eventually, after three rather sleepless nights on the maternity ward, in pain, with new-borns fussing and new mothers calling out for assistance in the night I was almost ready for home - save that my wound was oozing and I couldn't move the left side of my face.

I'd already mentioned to one of the registrars and several nurses that I thought my wound was infected, but I was told not to worry and that it was fine…….My husband had flown off on a business trip to Germany the previous day and so my mum had come to collect me from the hospital. Mum wasn't convinced that my wound was fine and neither was I - but I'd said my piece and I couldn't actually bear the thought of waiting around for another doctor to prod me …. The ward wasn't a great place to be and I longed to get my precious bundle home and to be there with my elder daughter who, at just twenty two months, needed me too.

Also, as I'd already mentioned my facial paralysis to the midwives (who were completely disinterested) I didn't see the point in asking to see a doctor about that either. In any event my mum, on looking at and examining me had already informed me that it was most likely a bell's palsy.

*"A bell's what?"* I thought to myself.

I had a vague recollection of a childhood friend having this and that it resolved eventually, but curious as ever I did some internet research when I got home. Rather helpfully the NHS website gives a good explanation which I've set out below:

*"**Bell's palsy** is a condition that causes temporary weakness or paralysis of the muscles in one side of the face. It is the most common cause of facial paralysis.*

*What are the symptoms?*

*The symptoms of bell's palsy vary from person to person. The weakness on one side of the face can be described as either:*

- *partial palsy, which is mild muscle weakness*
- *complete palsy, which is no movement at all (paralysis) – although this is rare*

*Bell's palsy can also affect the eyelid and mouth, making it difficult to close and open them and in rare cases, it can affect both sides of a person's face.*

*Bell's palsy is only diagnosed if other possible causes of these symptoms are ruled out, such as stroke or a tumour.*

*Why does it happen?*

*Bell's palsy is believed to occur when the nerve that controls the muscles in your face becomes compressed.* **The exact cause is unknown, although it's thought to be because the facial nerve becomes inflamed, possibly due to a viral infection.** *Variants of the herpes virus may be responsible.*

*Who is affected?*

*Bell's palsy is a rare condition that affects about 1 in 5,000 people a year. It more commonly affects those aged between 15 and 45, but people outside this age group can also suffer from the condition. Both men and women are affected equally.*

**Bell's palsy is more common in pregnant women and those with diabetes and HIV, for reasons that are not yet fully understood.***"*

So it's probably a virus that's more common in pregnancy ....Clearly, allowing me to suffer four weeks of contractions at home, a full labour followed by an emergency section had taken its toll on me and made me susceptible to catching this virus. I was already very cross that I wasn't allowed to bring my section date forward due to the traumatic birth I'd just had to endure but getting this condition on top of everything else was just the icing on the cake! (Or so I thought.)

But things continued to get worse with the facial paralysis and at day six post-delivery, during a home visit from a physiotherapist who was easing the tension in my exhausted body, she advised me to return to hospital as she was concerned that I could have suffered a stroke. I considered her advice and read the NHS guidance for myself and noted that bell's palsy is only diagnosed once other causes such as stroke and tumour are ruled out, so even though I was pretty sure I was OK and the last place I wanted to be was back in hospital I could see why my physiotherapist was concerned and was deliberating about whether I needed to speak to the hospital again……But thankfully during my deliberations I received a call from my GP who said that my physiotherapist had called her to express concern and that she would be making a home visit to see me that evening. When my GP arrived she felt that my symptoms were not conclusive of bell's palsy as she felt both sides of my lower face were affected so she called the medical registrar at A&E to arrange for my admission.

With a heavy heart I called my mum to tell her what had happened. I told her that I needed her to be at home with the girls whilst I went to hospital and because having two under two was a bit much for my sixty odd year old mother, I also called my maternity nurse to ask her to start work early.

Once they'd arrived I hugged my girls for dear life, handed my new born over to my maternity nurse, and with my breast pump packed, I got a taxi to A&E. Following an assessment I was sent straight to the emergency admissions ward. Thank god we could afford a maternity nurse was all I was thinking as I was moved on the trolley to the ward.

Basically, because I'd felt so poorly after the delivery I'd had to arrange for a maternity nurse to come and do some nights to help with my new-born, whilst my mum helped out with my toddler. I was also very grateful for my efficient breast pump. With my breasts larger than ever, tender and full of milk I didn't want to stop breast feeding, notwithstanding the fact I was about to be parted from my new baby girl.

Anyway, there I was at six days post-delivery, being separated from my toddler (again) and my new-born, being admitted as an emergency to the hospital I'd been discharged from only three days earlier. I was monitored on the emergency admissions ward and referred to a neurosurgeon. I lay there in my bed feeling incredibly alone and with my hormones crashing down post-delivery, the pain from my c-section wound and the sheer exhaustion of the labour and the preceding four weeks all weighing heavily on my mind, and on my shattered body, I wasn't feeling in a good place.

I cried myself to sleep that night.

I was missing my husband and my toddler, but most of all my precious new baby girl, with whom I'd been *"one"* for the preceding eight and a bit months. The ward sister was really sweet and called for a counsellor to come and talk to me as I was utterly inconsolable. I didn't want to talk though; I just wanted to cry and to rest.

Eventually I managed to sleep thanks to the zopiclone sleeping tablet I was given and thankfully, following a thorough examination by the consultant neurosurgeon the next morning, I was given some good news. He thought that I'd almost certainly developed a bell's palsy, but in order to be absolutely sure he wanted to confirm this with a brain scan.

I was given the option of remaining on the ward with continued monitoring and with the hope that I'd be able to be scanned in the next couple of days on the NHS (but there was no guarantee on timing) or returning home and arranging a private scan for two days' time. The choice was an easy one. I had private medical insurance and so being in the comfort of my own home with my girls, and a guaranteed time and date for a scan in two days, was a far more appealing option than staying for monitoring on an acute admissions ward with no guaranteed date or time for a scan and only my breast pump for company.

The brain scan I had two days later confirmed that I hadn't suffered a stroke and I was told that the facial paralysis was due to a bell's palsy and would eventually resolve. I had a grade III facial palsy and it took around six weeks for my face to completely return to normal.......

Meanwhile, my wound continued to ooze and when the midwife returned to my house for a home visit to remove the surgical staples a few days later she couldn't remove them all. Why? Well, because the wound was infected (of course) and the staples had become embedded within the infected wound. This was despite the fact I'd been repeatedly assured days earlier before my discharge that I didn't have a wound infection when I said that I thought I did (arghhh).

I therefore had to return to the hospital for the second time since my discharge to have the staples removed under local anaesthetic by an obstetric registrar. Which, I might add, was still incredibly painful as the local anaesthetic didn't numb the area completely and the local anaesthesia injection itself was in such close proximity to the infected wound that it just added to the pain… but the staples were at least removed.

A course of heavy duty antibiotics then followed.

Unfortunately, however, even when my wound did heal, things still didn't feel right with my tummy and I made numerous trips to the GP over the following few weeks with deep pelvic pain, vaginal discomfort and low back ache.

Then at six weeks post-delivery I went to my GP to get the coil fitted.

Whilst I was breast feeding (which has a natural contraceptive effect) I was also mixing my breast with bottle feeds, particularly on nights when I had my maternity nurse to help, and therefore I couldn't rely on breast feeding alone as a contraceptive. And whilst I was still exhausted from the birth and feeling pretty rank I had to consider my husband's needs too.....

Anyway when the GP inserted the speculum into my vagina it was absolute agony. My GP said that my cervix was severely inflamed, red and covered with pus. Needless to say I couldn't get the coil fitted and instead I was put on a combination of oral antibiotics to treat the infection, (metronidazole and ofloxacin). What upset me even more, however, than finding out that I had a rip roaring infection in my cervix (and undoubtedly throughout my womb) was that I was advised not to breast feed on these antibiotics as the effects on a lactating infant were not well understood.

This was a huge blow.

Not only was breast feeding better for my baby girl, it was also a special bonding time between us. I was really emotional about having to give up and I felt like I had been robbed. And whilst in theory I could have just expressed using my breast pump every time my daughter fed from the bottle (expressing the same number of ounces of milk as she drank and then disposing of it) the reality was that this was never going to happen. It would have been too time consuming when looking after my new-born and a toddler and when my baby could smell my milk she wanted to take my breast and not the bottle .... so the only real option was to give up.

So give up I did.

And stopping breast feeding quickly when your breasts are full of milk isn't easy or comfortable. In fact it is downright painful. To get the deed underway I put on my tightest fitting sports bra and gym top and left them in situ for two long days. This did the trick, and after forty eight hours or so without any nipple stimulation the milk supply had pretty much dried up.

During this time I was obviously in pain with my breasts, but more worryingly, my deep pelvic pain had become steadily worse since my visit to the GP. In addition, I'd started to throw high temperatures and broken out in a rash over my arms and legs (despite being on the antibiotics). I also had a crucifying headache and neck stiffness and couldn't tolerate bright lights. It struck me that by using the speculum in the surgery and attempting to insert the coil my GP had actually made the infection worse. Once again my husband was away on a business trip and so I had to call my mum to come over and help me with my girls. I was simply too poorly to look after them alone.

On arrival my mum took one look at me and said I looked like death (and in truth I wasn't feeling much better either). She was concerned that I had meningitis or sepsis symptoms and thought I needed to go to hospital. I was on high doses of oral antibiotics but was still throwing a fever and was getting progressively more poorly, not better. As I was feeling too ill to drive and my mum was trying to balance looking after both my new born and my toddler I decided to call an ambulance.

I felt so low and so poorly.

I hadn't felt properly well since my threatened pre-term labour which was by now ten weeks ago. I had a strong sense of injustice because six weeks after a caesarean I should have been feeling good and enjoying my new baby girl.

But I wasn't.

I was about to get into an ambulance and be rushed to hospital……..I was also very angry at my consultant obstetrician (the short, fat, balding, arrogant one) because I couldn't help feeling that the physical stress of being left to suffer contractions for four weeks and then being persuaded to try for a vaginal delivery were the root cause of my post-delivery complications. I knew I was going to labour early, but my consultant point blank refused to bring my section date forward. He didn't listen to me and didn't take into account that I'd delivered a 7llb 1oz baby at thirty seven weeks first time around....

He was dismissive in his dealings with me and as far as I was concerned he made a grave error in judgment...... (But what did that matter to him? He wasn't the one suffering.....)

Anyway, on arrival at the hospital I was assessed in A+E and given some pain relief and then switched to IV antibiotics. As my rash was described as being a *"blanching"* rash the medics were sufficiently concerned about the possibility of meningitis that my consent was sought for a lumbar puncture to be performed.

Of course I gave my consent.

I felt so terribly poorly and I just wanted them to do anything to get me better, even though it'd been explained to me that a lumbar puncture is not a risk free procedure. There is a chance of damaging the spinal cord (which carries with it a chance of paralysis).

Thankfully the lumbar puncture revealed that I didn't have meningitis, but because I was still so very poorly I was kept in hospital for four days on antibiotic treatment until my symptoms started to ease. I was then transferred back to oral medication and allowed home.

I wasn't actually given a reason for the rash or the headaches, stiff neck and photophobia, but was just told that I had an infection of the genital tract following delivery, some pelvic infection and that the wound infection may also have contributed to my symptoms.

My C - reactive protein (**CRP**), a marker for inflammation, was raised which indicated a bacterial infection and physically I felt that my body was just totally spent. After all, I'd recently endured four weeks of contractions, a full labour, an emergency caesarean, a bell's palsy, a wound infection and a severe genital tract and pelvic infection.

Whilst an in-patient during this latest admission I specifically requested to see the consultant obstetrician, under whose care I should have been during my delivery and with whom I had paid privately to have all my ante natal scans. But as he didn't come when I requested I walked around to the ante natal ward one afternoon, cannula in situ, and asked the lady on reception whether he was in clinic that afternoon. When she remarked that he was I said in a loud and angry voice:

*"Well can you please ask him to come and see me about my genital tract and pelvic infection, which I wouldn't be suffering with right now if he'd bothered to give me the right care before my labour ...... I have a six week old baby at home who I can't breastfeed because of my infection and I'd be grateful if he would at least have the courtesy to come and see me rather than fob me off with junior doctors.....I feel like death and am getting no better."*

I didn't wait for her to respond and by the time I'd finished my little tirade (which became more husky as I went on because I was so fatigued) I had tears streaming down my face.

My visit worked though because he did eventually come to visit later that day. But he was incredibly rude and arrogant, standing briefly at the end of my bed, staying just long enough to talk at me, telling me very abruptly that there was nothing to worry about and that everything would clear up with the antibiotics. He didn't even stay long enough for me to open my mouth to say anything at all let alone for me to relay my concerns over the decisions he made prior to my labour and the impact these decisions had had on my current situation.

And whilst I could have followed up with some sort of complaint, I didn't really have the energy or see the point at the time. Also, by the time I was discharged I just wanted to get on with looking after my new arrival rather than waste time on negative things. It was, however, quite apparent that things weren't still quite right with my health. In particular, my deep pelvic pain persisted and I was beginning to think that perhaps something had been left inside me during the caesarean.

In fact, by the time nine weeks had elapsed since the birth, I was still suffering with intolerable pelvic pain and so I took myself off to my local private hospital Emergency Care Centre (**ECC**) where I hoped I could get a quick referral to a consultant gynaecologist. High vaginal swabs were taken by the nurse in casualty and when I returned a couple of days later to see the consultant gynaecologist, he confirmed that I still had some bacterial vaginosis despite all the antibiotic treatment I'd had. So, I was given a six week course of metronidazole gel and told to come back after the course had finished for a review of my symptoms.

I also relayed my concerns about my deep pelvic pain to my new private gynaecologist and he listened intently (unlike his contemporary who'd fobbed me off during my pregnancy and on the ward three weeks earlier). My new gynaecologist expressed disbelief when I recounted to him about being left contracting for four weeks and then being persuaded to try for a vaginal delivery when my birth plan was clear about me having an elective caesarean. (As an aside this gynaecologist has three children of his own and had his wife deliver all three by planned caesarean so as to reduce the risk of birth trauma as much as possible...... Clearly he was all too aware of the appalling state of our midwifery led childbirth service and was taking no chances with his loved ones.)

During this appointment I also recounted to my gynaecologist the incident shortly after the birth whilst I was recovering in a side room when my husband showed me the pictures he'd taken hours earlier in the operating theatre ….. What my husband showed me was a picture taken immediately prior to our baby being lifted out of my abdomen in which you could clearly see my small intestine laying outside of my body, to the left, on the operating table. I have to say I felt incredibly nauseous when I saw the photo, but didn't think too much of it until later that day when my father visited.

My husband showed my mother and father the same pictures and my father was quick to remark that he couldn't fathom what the obstetrician was doing in my peritoneal cavity and what my small intestine was doing laying besides me – he said he'd be surprised if my bowels would still be working OK after this…….

Whilst my gynaecologist didn't say too much, it was evident that he was appalled. He also asked me whether my bowels were working, to which I said yes, but that I had been a little bit constipated and had a bit more wind. His response was simply that this often happens after surgery and that he felt things should settle down but that I should come back earlier than the follow up appointment scheduled for six weeks if things got any worse.

Anyway, during the six week course of metronidazole gel my periods resumed and when they did they were excruciatingly painful and extraordinarily heavy. In fact just prior to my first menstruation I felt this ripping sensation in my abdomen which was almost as painful as full blown labour contractions. In addition, my deep pelvic pain had started to worsen rather than improve and I suffered pain on intimacy with my husband.

So I returned to see my gynaecologist before the six weeks were up and I explained that I was feeling worse since my periods returned. We discussed my symptoms and he advised me that I needed to have an investigative hysteroscopy and laparoscopy to assess what was happening inside and given the recurrent infections I had endured he wanted to rule out there being any retained products. And whilst no one relishes the thought of an operation I was actually quite pleased that I was going to have the laparoscopy because I'd been so poorly since the birth.

At the laparoscopy my gynaecologist found some bowel attached to the under surface of my abdomen, adhesions over the surface of my uterus and bladder, my right fallopian tube twisted back on itself and stuck to my bladder and several small spots of endometriosis. My gynaecologist also remarked that my uterus was soft and bulky and he suspected that I had adenomyosis, which could also have been the cause of the dyspareunia (painful sexual intercourse).

When I discussed the findings of the operation with my father he was quick to remark that since I'd never suffered with endometriosis before my daughter's birth, the obstetrician who delivered my now four and a half month old second daughter, was no doubt as careless with my endometrium as he was with my bowels, distributing bits around my abdominal cavity when closing me up......Certainly the operation had been fraught and was carried out by a registrar and not a consultant and I have absolutely no doubt in my mind that the endometriosis was as a result of the caesarean.

I had no trouble before this birth with endometriosis or painful periods or painful sexual intercourse.....and to me it was as plain as the light of day. Further, even a cursory internet search of decent medical literature confirms the fact that it is not uncommon for endometriosis to first appear after a caesarean section.....

# Loss of my womanhood at thirty three

I decided to return to work as a lawyer on a part time basis shortly after the laparoscopy and hysteroscopy, when my new arrival was only five months' old. I had intended to take up to a year's leave, but because I'd been so poorly during these early months I felt that returning to work would give me more of a sense of normality. I also hoped that any illness following the birth of my second daughter would be at an end and that working two days in the office and one from home would give me a good balance in life.

But unfortunately I didn't find that balance because my ill health was far from over. I still felt very poorly. My periods continued to be excruciatingly painful with me feeling like my insides were being ripped out each time my womb was shedding its lining, sexual intercourse became increasingly more painful and my bleeding increasingly heavy. I also started to feel more constipated and was starting to pass unpleasant smelling wind. And to add insult to injury I was finding it harder to climax during intercourse, because in addition to the pain I felt during deep penetration, my clitoris was far less sensitive.

I certainly had none of these problems before the birth of my second daughter the previous September and by the following March I was fed up with the pain, the bleeding and the interference with my sex life, that at thirty three years of age, I discussed the possibility of a hysterectomy with my gynaecologist. He explained to me that I was almost certainly suffering from a condition known as adenomyosis and that the only cure was a hysterectomy. He noted that my uterus had felt bulky during the previous laparoscopy and if the pain was too much to bear and I wanted no more children a hysterectomy could be a good option for me.

I explained that I didn't want to menopause early or lose my cervix as I felt this was important for mine and my husband's sexual enjoyment and I asked if there were ways around this. My gynaecologist explained that he could leave my ovaries, fallopian tubes and cervix behind and that I wouldn't necessarily menopause early if he did this. He explained that I certainly wouldn't menopause straight away but it may bring my menopause forward by a few years. This seemed like a reasonable prospect, particularly if it cured my deep pelvic pain and the pain on menstruation and during sexual intercourse, and so in the April before my thirty fourth birthday, almost eight months after the birth of my second child I, opted for a sub-total hysterectomy to rid me of the adenomyosis.

The Mayo Clinic website gives a very good definition of adenomyosis and also offers information on the causes:

*"Adenomyosis (ad-uh-no-my-O-sis) occurs when endometrial tissue, which normally lines the uterus, exists within and grows into the muscular wall of the uterus. This happens most often late in your childbearing years after having children…….*

*For women who experience severe discomfort from adenomyosis, certain treatments can help, but hysterectomy is the only cure. …..*

*The cause of adenomyosis isn't known. Expert theories about a possible cause include:*

***Invasive tissue growth.*** *Some experts believe that adenomyosis results from the direct invasion of endometrial cells from the surface of the uterus into the muscle that forms the uterine walls. Uterine incisions made during an operation such as a* ***caesarean section (C-section)*** *may promote the direct invasion of the endometrial cells into the wall of the uterus.*

***Developmental origins.*** *Other experts speculate that adenomyosis originates within the uterine muscle from endometrial tissue deposited there when the uterus first formed in the female foetus.*

*Uterine inflammation related to childbirth.* Another theory suggests a link between adenomyosis and childbirth. An inflammation of the uterine lining during the postpartum period might cause a break in the normal boundary of cells that line the uterus.

*Stem cell origins.* A recent theory proposes that bone marrow stem cells may invade the uterine muscle, causing adenomyosis."

On reading this I was absolutely certain that the adenomyosis was also an unwanted result of my recent caesarean. And whilst I was obviously upset at losing my womb and thus my ability to have any more children, my overwhelming feeling about my impending loss of my womanhood was one of anger.

Anger, because I instinctively knew that had I been given my caesarean section a week earlier, as I'd requested, then all of this would have been avoided. Whilst I acknowledged that the National Institute for Health and Clinical Excellence (**NICE**) guidelines stated that planned caesareans should not be routinely carried out before thirty nine weeks I knew that my case was different because of the amount of time I'd been experiencing painful contractions. I was upset that by not listening to me and by not looking at my case on an individual level, taking all factors into account, a sensible decision wasn't taken which could, in my opinion, have avoided my further surgeries.

In particular, I'd never suffered from endometriosis before, nor did I have painful and heavy periods and so the most likely cause of the adenomyosis was the caesarean. Clearly, having an emergency caesarean where my baby was impacted and had to be pushed up my birth canal whilst my abdomen was open caused significant trauma. Had I had a planned caesarean at an earlier date, and prior to me labouring naturally, this trauma would have been avoided.

On the positive side, the operation for the hysterectomy thankfully ended the heavy bleeding and excruciatingly painful periods and lessened some of the pain on intercourse, but it didn't eradicate all of the pain on penetration and my whole abdominal area still didn't feel right. I was also still suffering with an element of sexual dysfunction with a loss of some sensation in my clitoris, making it much harder to achieve orgasm. And in addition to this I started to experience night sweats and suffer with a more sluggish bowel, nausea and increased wind.

However, after several further trips to see my gynaecologist and GP over the subsequent eighteen months I was given no clear diagnosis or reason for these symptoms and so I eventually resigned myself to the fact that nothing else could be done for my intermittent abdominal cramping. That was until I started to become increasingly poorly.

# No, something is wrong ……..

My symptoms of constipation, excessive wind, night sweats, abdominal cramping and painful intercourse continued to worsen until I reached crisis point.

It was the December following my youngest daughter's third birthday. I came down with a flu-like virus with aches and pains all over my body shortly before Christmas. I then had some vomiting and diarrhoea, for a couple of days, including on Christmas day. I drank and ate very little for around two weeks and was so weak that I spent most of that time in bed. Then as I was recovering I realised that I'd not been for a bowel movement for the whole time I'd been poorly (apart from the two or so days of diarrhoea) so I took some lactulose to ease the constipation (and that seemed to help). I then returned to work (as a partner in a small boutique law firm) shortly after new-year on a Tuesday, feeling much better but still not brilliant.

By this point in time (due to various financial pressures in a failing economy) I'd upped my working schedule to four days in the office. (Although in reality I was cramming a full working week into those four days whilst taking roughly sixty percent of a male equivalent's full time wage). Anyway, as I still felt bunged up, even after the lactulose, I self-administered some senna tablets that week on the Thursday evening, after work, to see if that would help get things moving a bit more.

The senna appeared to work as I managed to open my bowels really well the following morning (which was my day off), but by mid-afternoon that same day I started to get agonising contraction type pains in my abdomen and a strong urge to defecate. I was in my gym and country club in the lounge enjoying a cappuccino, whilst on a teleconference call with another solicitor and barrister when the pain first started..... I had to end the call early because I couldn't cope with the pain and needed to get to the toilet quickly. Yet when I went to the toilet I couldn't actually defecate and instead I was just sitting there in absolute agony as if everything was stuck. The pain was as powerful and as toe curling as full blown labour.

As I got up from the toilet the pain eased slightly but it quickly became apparent to me that I would be unable to collect my girls (my three year old was in the crèche at the club and my five year old at school) or drive anywhere and that my husband would have to leave work to help out. So I called my husband and he left work early to collect the girls.....

Meanwhile I was doubled over in the foyer of my gym. The pain was colicky and griping and came in waves. It was rapidly becoming too much for me to bear without any pain relief and I began to scream out in pain. As I was screaming out in pain the club manager came to see if I was OK but she quickly realised from the expression on my face and the way I was holding myself that I wasn't and so she called an ambulance. It felt like an age for the ambulance and paramedics to arrive, but when they did, I was told that they'd only taken just over ten minutes.

On arrival the paramedic quickly assessed me, gave me gas and air straight away, cannulated me on site and gave me a shot of IV morphine. Thankfully this took the edge off the pain and I was able to stop screaming, and remain relatively calm as I was pushed in a wheelchair to the ambulance.

On arrival at A+E (at the same hospital where I'd given birth to my girls) the pain relief was continued and I was also given antiemetics as I felt like I was going to vomit. I had a rectal examination (which involved a doctor sticking his finger up my backside and wriggling it around) and a plain abdominal x-ray. What the rectal exam revealed was some hard pellet like faeces in my rectum and the x-ray showed some faecal loading.

So I was admitted to a surgical ward overnight, given some IV fluids, more pain relief and anti-emetics and the following day I was given a microlax enema. The enema had a reasonable effect and I managed to have a bowel movement, but it didn't feel like much so I presumed that constipation couldn't have been causing my pain.

Anyway, as the morning went on I was starting to feel better so I asked to go home. It was a Saturday and I really couldn't bear the thought of spending the weekend in hospital, where not much happens in terms of tests or diagnosis or indeed consultant visits. However, as I still felt quite poorly over the remainder of the weekend I decided to work from home the following week to give me time to recover from whatever it was that gave me such pain the previous Friday. And I was glad that I did because by the Wednesday I started to feel nauseous again and my abdominal cramps returned to the point where I could no longer bear the pain.....

So I took myself back to the same A+E department where a further abdominal x-ray was taken which still showed faecal loading - despite the enema I'd been given the previous Saturday. I was obviously keen for them to find out what was causing my pain and constipation as it hadn't resolved so I stayed for them to carry out an ultrasound of my abdomen, pelvis and kidneys.

I ended up staying in hospital for five nights and six days, during which time the only diagnostic tests carried out were an x-ray and an ultrasound. The surgeons couldn't decide what was wrong with me and so referred me for a gynaecology opinion, but on review the gynaecologist decided whatever was causing my pain was not gynaecological. The only diagnostic information the doctors had from this stay was my own patient history, an ultrasound which revealed a cyst on my right ovary and the x-ray showing faecal loading.

Both the symptoms I described and the x-ray suggested constipation and I was therefore given some picolax (a very strong laxative, used primarily as a bowel preparation prior to investigations or surgery of the bowel).

Well, the picolax was a revelation! .... And a really unpleasant one at that!

I had to drink two lots of foul tasting liquid several hours apart and shortly after taking the second dose I was in absolute agony. I could feel the liquid starting to work its way through my bowels and it felt how I imagine paint stripper would feel like to paint.

I felt an excruciating *"ripping"* sensation on the left side of my abdomen as the laxative was taking effect. It was one of the most painful episodes I have ever had to endure and it felt as if my bowels were tearing away from my abdominal wall. (Which in fact they probably were.) Then when my bowels started to open I literally didn't move from the nearest ward toilet for what seemed like hours.

On asking for more morphine whilst the picolax was taking effect I did relay to the nurses the fact that I was in complete agony and that it felt like my bowels were tearing away from my abdominal wall (which to a mere lay person seemed quite significant) but the nurses were quite disinterested. In fact when going through my hospital notes at a later date whilst contemplating litigation I also realised that there is not even any entry about this episode of acute pain. So, even if a doctor reviewing my care were to consider this of any significance they would be prevented from doing so due to its absence.....

On my discharge from hospital I felt very bewildered that I had been admitted to hospital for six days, but didn't receive any clear idea from any of the medics what could be causing the constipation and pain. I did, however, start to take laxatives on a more regular basis following my discharge and decided to wait for my follow up appointment with the consultant in two months' time to see if I would get any clearer idea of what was going on.

Unfortunately, however, following my discharge I continued to feel nauseous and bloated and my night sweats increased in their frequency, then about a month after my last discharge, whilst I was at my mother and father's house I started getting horrific colicky cramps again, but this time with uncontrollable vomiting. I was literally vomiting nonstop. It was unlike any vomiting I'd previously experienced. I would vomit, but before I had a chance to recover properly from the vomit to catch my breath I would just vomit again.

Eventually there was nothing left to vomit, not even bile but still then I continued to retch from deep in my abdomen, heaving with each retch and doubling over with the pain. I had to start taking small sips of water when I could because I was so overcome with an unquenchable thirst due to the extensive vomiting. Yet all that happened when I drank was that I would vomit up the water again almost immediately. This went on for several hours until the pain from the cramping, griping, vomiting and dry reaching was too much to bear.... at which point my husband took me to the A+E department local to my parents' house.

I was relieved in a way that I'd fallen ill at my parents this time because their nearest A+E was a large and reputable university teaching hospital. (My local hospital is a peripheral district hospital which arguably doesn't attract the best doctors...) I was also seen quite quickly by the doctors to be cannulated as it was readily apparent to the staff that I was in agony and couldn't keep even the smallest sips of water down, let alone the codeine they gave me shortly after being triaged. So as soon as my cannula was in I was given IV morphine and antiemetics. The antiemetics worked quite quickly and I was given several bags of fluids over-night and was admitted to a surgical emergencies ward.

Whilst an inpatient I had a gastroscopy and an abdominal ultrasound, neither of which showed anything of note and I also had a follow up CT scan of my upper gastrointestinal tract as an outpatient. And despite me telling the doctors of my earlier admissions where faecal loading had been apparent on x-ray no one actually ordered a plain abdominal x-ray for me (which almost certainly would have revealed faecal loading again)......Perhaps more worrying though, the CT scan remarked that I had:

*"a slightly bulky uterus but no discreet abnormality"*

and

*"no abnormality is seen in the unprepared colon"*.

As I'd had a hysterectomy three years previously quite how the scan revealed a phantom uterus I have no idea! (I only discovered the reference to the phantom organ when I asked for my clinical notes from my lawyers some years later). And as for *"no abnormality"* in my colon – I will deal with this later on......

Anyway, following my discharge from the university teaching hospital but before my outpatient follow up I had a follow up appointment in the clinic of my local hospital.

I attended the appointment hoping that given my further recent admission (which no doubt they'd know about by now) I'd be given appointments for a battery of further tests....

I waited an hour to be seen and then when I eventually saw the registrar who was covering for the consultant, I asked if he was aware of my further recent admission. He wasn't. But not only was he not aware of it, when he found out it was at a different hospital he simply responded by saying that I could be discharged from his clinic as I would be followed up there!

I was so incensed by his abrupt and dismissive attitude that I didn't push back and insist on a proper consultation. It was blatantly obvious that he just wanted to reduce the number of patients in his clinic and didn't want to listen to me at all so I didn't bother wasting my energy. Then by the time I was followed up in the clinic of the teaching hospital, I was taking laxatives on a daily basis to ensure I was moving my bowels regularly and I was starting to feel a bit better. So, when I was told that the CT scan was clear and that it was unlikely that there was anything further down my gastrointestinal tract that would have caused the vomiting, I rather naively accepted the explanation and was content to be discharged. Again, this was despite me recounting the fact that previous x-rays had showed faecal loading and that I was having to take regular laxatives.

Looking back now I'm angry that none of the doctors I saw thought that my constipation was worth investigating.

Needless to say, following my discharge from both hospitals my night sweats continued to increase in frequency and severity and I was becoming increasingly constipated. With nowhere to turn I visited my GP. My GP was disinterested in the constipation and merely remarked that *"lots of women suffer from constipation after childbirth"* and she simply suggested that my night sweats were almost certainly due to early menopause. She suggested putting me on hormone replacement therapy (**HRT**) but first said that I needed a follicle stimulating hormone test. After one test I was called back and put on HRT, but I've since discovered that two tests are required to prove menopause...... Unsurprisingly the HRT didn't stop the night sweats and the total lack of interest in my constipation symptoms, notwithstanding my three emergency admissions in agony earlier in the year, frustrated me.

Then in the May of that year, the day before my thirty seventh birthday I was admitted once more to my local hospital with severe nausea and sudden onset colicky abdominal pain. The pain was so excruciating that I had to receive quite a high dose of IV morphine in A+E to get my symptoms under control.

As before a plain abdominal x-ray showed faecal loading.

Two things really stand out for me from this admission. The first is my girls and my husband bringing my birthday cake into the hospital canteen where we lit the candles, sang *"happy birthday"*, ate the cake and opened presents! The second thing that stands out is the total and utter disregard with which the medical staff treated me.

In particular, one registrar told me that there was nothing wrong with me and that I should just go home and get on with my life.

He said:

*"Just because I might have a pain in my leg I don't go to hospital and likewise you don't need to be here with your pain. There is nothing we can do as the pain is probably from adhesions from previous surgery and any further operations will just cause you more problems. You will just have to live with the pain."*

I was totally speechless when this doctor said this to me. Did he really think that I would be in hospital away from my husband and young children if there was nothing wrong with me?

I'd been admitted in excruciating pain, with nausea and drenching sweats and there was no way I could deal with the pain at home, yet he spoke to me as if I was a complete hypochondriac. Then to make matters worse he enquired about my mental health and asked if I was anxious or depressed thereby intimating that any physical health issues were all in my head. This made me furious beyond belief. Of course I was a little anxious because I was having repeated hospitalisations and despite tests no one could come up with a diagnosis, but aside from that I was totally fine on the mental health front!

What angered me most was that he didn't appear to listen to a word I'd said about the onset of the colicky pain and the history of the problem from the start of the year, and instead he just dismissed me as some kind of fruitcake.

I can't remember exactly what I said in retort to this man, but I can remember raising my voice and swearing, not at him, but out of frustration and in an effort to stress how awful I felt. And when I read through my notes at a later date there is an entry about me swearing and behaving in an aggressive way! I actually found this almost comical because there are plenty of important clinical omissions throughout my notes, yet someone had bothered to make an entry about my behaviour which was borne entirely from the pain, frustration and nausea.

The investigations carried out during this admission were again an x-ray showing faecal loading and an ultrasound revealing the cyst on my right ovary. I was once more referred by the surgical team for a further gynaecology opinion. I obviously discussed with the gynaecologists my previous gynaecological surgery and rather unbelievably the advice I was given by the gynaecology registrar was to get re-referred privately to the gynaecologist who undertook my laparoscopy and subtotal hysterectomy (privately) because:

*"he knew what I looked like on the inside"!*

Really?!? Like he'd actually remember?!

There was no attempt at all to address why I could be getting constipated and why I had such excruciating pain and it was quite obvious to me that the doctors who I'd seen had no interest whatsoever in trying to find out the cause of my current health problems. During my inpatient stay I'd been given enemas to help clear me out which was helping to treat the problem, but no one had attempted to address the cause of the constipation. And, of course, I'd been told that there was nothing wrong with me anyway.....

Believe me, when you have been admitted to hospital four times in less than six months in excruciating pain for stays ranging from two to ten days and the doctors tell you that the investigations reveal that there is nothing really wrong, you do indeed start to question your sanity.....But only momentarily...... Because the physical symptoms were so strong and so real that I didn't ever let the questioning of my mental health deter me from trying to find out the cause of my sweats, my constipation and my horrific pain.

So immediately upon my discharge I booked a private appointment with the gynaecologist who had performed my previous laparoscopy and my hysterectomy and because all my x-rays were showing faecal loading and no one had given me a reason for this I also wanted to see a general and colorectal surgeon. I therefore also booked a private appointment to see one.

I didn't even bother going via my GP for a referral because it would have taken too much time and as I was self-paying rather than insured at the time I didn't have to go via my GP for insurance purposes. The colorectal surgeon I elected was one of the consultants I'd very briefly seen the previous Saturday on his ward round at my local NHS hospital – but because I was under a different consultant he couldn't really treat me. (Quite what the point was of him even seeing me when I was under one of his colleagues and couldn't be treated by him I have no idea ….) So I simply looked up his details on the internet on discharge and booked in with him via his secretary.

In terms of timing, I was discharged from the NHS hospital on the Wednesday and saw the consultant general and colorectal surgeon privately on the Friday. Unfortunately, however, I came away from the appointment feeling really disappointed because I wasn't given any idea why I could be getting the colicky pain or why I was constipated. He arranged no investigative tests and just told me to take some laxatives for a few weeks and then to come back and see him.

But because I'd been so poorly for over six months by this point, and because my bad health was adversely affecting all areas of my life, I wanted some proper investigations carried out as soon as possible. I therefore explained to him that I'd already been taking laxatives and that I was still getting these episodes and my x-rays were repeatedly showing faecal loading.

Notwithstanding relaying this information he still didn't seem to want to do anything other than send me away with the laxatives and then review me in a few weeks - and this simply wasn't good enough for me. I really needed someone to take my symptoms more seriously and propose a proper action plan before my health (and life) deteriorated further.

In fact a large part of my desire to get a swift resolution to the problem and get back on my feet as quickly as possible was that my health was hugely interfering with my work. I'd only joined my law firm as a salaried partner the previous September and had missed approximately six weeks off work due to these recurrent episodes requiring hospitalisations. I didn't have any employment rights and as we were in the midst of the "*great recession*" I knew that patience was wearing thin. Moreover, we'd just had a family business enter administration and as my husband was in the early stages of building a new business he wasn't actually drawing any income. I was therefore the sole income earner in the family at that time.

So when I went to see my gynaecologist the following Monday and he was immediately more empathetic and wanted to do everything he could to help me I instinctively trusted his judgment more than the colorectal surgeon and felt that he was the right person to help me.

My gynaecologist scanned me that evening and said that he could see the cyst on my right ovary and also one in my right fallopian tube and some shadowing on the scan. He was totally honest with me and said that he didn't know whether this was causing my pain or if it was my bowels but that he could remove the cysts together with the ovary and tube and we could then see if this would make a difference to my pain. He also said that it's sometimes possible to get a rumbling appendix and as a lot of the colicky pain was on the right side of my abdomen he could perform a laparoscopy for the removal of my ovary and fallopian tube and remove the appendix at the same time.

Because I was being given a treatment plan I really didn't care that there was no certainty that the proposed surgery could cure my pain, or that I would be losing an ovary, I was just so relieved that my gynaecologist was prepared to do something to try and help me. Also, what he proposed seemed sensible so I agreed.

My operation was scheduled for ten days' time but I woke up two days after my consultation with acute colicky pain once more and so was admitted as an emergency to my gynaecologist's NHS hospital and had the operation two days later.

I was given picolax as a bowel preparation the day before my surgery, which again left me feeling drained and in agonising pain. In fact I felt the excruciating ripping sensation once more when it was flushing through my bowels.

At surgery my right ovary, right fallopian tube and appendix were removed and I also had some endometriosis resected and an adhesion removed which was between my bowel on the left and the top of my cervix. (I've subsequently been told by a different gynaecologist that there was absolutely no need at all to remove the ovary as it was only a dermoid cyst..... But I'll not dwell on this here.)

Anyway, following this operation I initially felt a little improvement. With hindsight that was almost certainly more to do with the fact that the picolax I took before my procedure completely emptied me out. I also continued to take regular laxatives, but by the November of that year, so less than six months after my surgery, I started suffering again with severe cramping pain and drenching night sweats. I needed to see a doctor, but didn't see much point in going to my GP as she'd been totally useless when I'd previously visited with my symptoms, and as my gynaecologist was the only person who'd actually really listened to me it seemed sensible to go back to him.

At my appointment my gynaecologist felt that my left sided pain could be to do with my remaining left ovary or possibly my bowels.

He told me to keep a diary of my symptoms and bowel movements and then to go back and see him in three months' time.

I wasn't convinced about the pain coming from my ovary and as I'd not yet menopaused I didn't really want to lose the last vestige of my internal womanhood and deal with all the crap that goes with it! Plus, as time passed, and with the help of the symptom and bowel movement diary I began to realise that the operation I'd had that summer (whilst not being completely unnecessary because I had two cysts and my appendix removed, some endometriosis ablated and an adhesion resected) was almost useless in terms of the relief of my symptoms.

So, by the time the February came and I was due to return to see my gynaecologist, keeping a diary had pretty much crystallised things in my mind.

I was becoming increasingly constipated and as the constipation worsened so did my nausea, cramping and night sweats. I was therefore absolutely certain that my cramping and nausea were bowel related. I therefore didn't return to see my gynaecologist and instead went to see my GP about my constipation and asked for a referral. My GP, once again, however, was totally useless and just told me to keep taking the laxatives and prescribed me something for my nausea. She didn't appreciate the extent to which my constipation was actually making me ill.

In fact it was making me so ill that I was toxic and was sweating out sulphur!

I knew this because my sterling silver necklace from Tiffany & Co kept going black after each clean. This started to happen just before the Christmas following my last surgery and nothing I did would make it stay clean. I therefore thought there was a problem with the silver necklace itself so I took it back to Tiffany & Co to see if they could explain it to me. The manager said that sometimes a change in perfume or chemicals in the environment can set off tarnishing on silver and rather than offer to replace it she said she would have it cleaned and asked if I would pop back and collect it later that day. So I went back and collected my necklace, but again, within a couple of days, it became very tarnished. So I returned once more to the store, at which point the manger sent it away to be cleaned by a special machine.

Even after this clean, however, the necklace became tarnished again after a few days wear and so I demanded a replacement necklace, notwithstanding the fact that the manager did all she could to assure me that there was nothing wrong with the necklace and that the silver must be reacting to something on my skin.

Being a lawyer, I threw the Sale of Goods Act at her and said that the necklace, quite simply, was not of satisfactory quality. Not wanting a stand up row in the shop, the Manager duly gave me a brand new replacement necklace. So confident was I that the necklace was not satisfactory that I assured her that if the new necklace tarnished I would give her the new necklace back in exchange for my old one! But when the brand new necklace became tarnished after only a couple of days wear I was just far too embarrassed to return to the store ……….

But it was actually this tarnishing of the new necklace and what the shop manager had said to me that made me realise that it had to be something that my body was producing which was reacting with the necklace to make it tarnish; I had worn the same perfume for the best part of fifteen years and was not being exposed to any other new chemicals or sprays.

So I did some internet research and noted that sulphur tarnishes sterling silver. Sulphur gasses are also produced in the colon by the friendly bacteria breaking down foodstuffs. *"BINGO!!"* I was clearly so constipated that I was sweating out toxins, including sulphur.

By that March, the realisation that I was sweating out sulphur, coupled with my increasing nausea and exhaustion from feeling poorly, spurred me into booking another appointment with the colorectal surgeon I had seen the previous June. (Even though on that visit he had just sent me away with some laxatives....)

On this visit he was, however, much more prepared to take action and he immediately increased my movicol dosage to four sachets in a litre of water at 2pm each day and asked me to write and tell him what happened when I increased the dose. More importantly though, he arranged for me to have some colon transit studies performed and a colonoscopy with biopsies.

My email updating my surgeon on what happened when I increased the dose went like this:

*"I took 4x sachets of movicol in a pint of water at approximately 2pm. This worked well and I had a number of bowel movements throughout the afternoon/early evening. The stools were generally normal looking but some were thin and bent/curved. I had an early night (8:30pm) as I felt very tired and weak and had a bad headache. The left side of my stomach[sic] (my sigmoid colon I think) was very painful and at times the pain became quite acute when there was any wind or motion movement in that area. I woke up twice in the night, both times in a drenching sweat, to have further bowel movements and my abdominal pain continued.*

*I drank at least 2.5 litres of water throughout the following day and evening (and have done this each day since) and then on the third day of increasing the dose I woke up feeling tired and weak and with considerable pain in my left side. The pain since having emptied my bowels following the Movicol the previous day was far worse than before my bowels had been emptied - it is as if the motions passing through that area of my colon have aggravated it further causing the pain. At one point it felt as though the area on the left through which the stool was moving was ripping my insides, which was excruciatingly painful. (I have also previously experienced this ripping sensation on the left of my abdomen when taking picolax.) Anyway, I went back to bed for a further sleep mid-*

*morning (about 10am) as I felt unwell and woke from this sleep again (about 11am) in a drenching sweat needing a further bowel movement. My sigmoid colon pain continued to worsen and I started alternating between ibuprofen and paracetamol to keep on top of the pain.*

*Whilst I had already experienced a considerable amount of bowel movement I continued with the treatment plan as directed and took a further 4 sachets of movicol in a pint of water at approximately 2pm. As the afternoon progressed the pain started to ease until I felt much better in the early evening. I didn't have any further bowel movements until around 8pm and by this time my movements were not well formed (but not loose either) but were short and thin and plentiful in number.*

*That night I woke up once for a bowel movement which again was plentiful in number and short and thin. By the morning I was feeling much better again and woke only in a light sweat rather than the drenching sweats I had experienced for the previous few days. I had two further movements the next morning which were the same again and a further movement that afternoon before taking a further 4 sachets of movicol mid-afternoon. By this point I was relatively pain free in terms of the ripping pain in my sigmoid colon but I was still aware of the pain - more like a constant dull ache in that area.*

*The next morning I woke only in a light sweat but my sigmoid colon pain had worsened (albeit not to the level it was when I first increased the dose) and I had four movements that morning which were thin and short and plentiful and I have been passing foul smelling wind.*

*Taking the four sachets is definitely getting things moving but equally it makes it difficult for me to lead a normal life/attend work due to the frequency of the movements and also the amount of wind I am passing……"*

Basically the increased laxatives were having a good effect and I was starting to feel better because I was emptying out more, but I was constantly on the toilet and passing foul smelling wind and so I couldn't really leave the house.....

For the colon transit studies I had to swallow a capsule containing twenty radiopaque markers which show up on x-ray. X-rays were then taken at three, four and five days following the ingestion of the capsule. In a normal healthy person it usually takes around forty two to seventy two hours or two to three days for the transit of waste through the colon and for all the markers to be expelled. My x-rays were, however, showing seventeen of the markers still in the colon after three days, fourteen after four days and thirteen markers still remaining in my colon after five days, or one hundred and twenty hours. This was despite me continuing to take laxatives throughout the duration of the x-rays. (Clearly I was not just suffering with normal constipation as my GP had been telling me.)

Indeed when I met up with my consultant to go through the results of the transit studies he remarked that I had significantly delayed colonic transit and that this was almost certainly what was making me feel so poorly.

Then about a week or so after my transit studies were undertaken I had the colonoscopy, which is basically a camera investigation of the colon. Once again I had to take the dreaded picolax so that my bowels were empty for the procedure. As before, I was in agony as the bowel cleanser worked. For the procedure itself I had IV sedation, but I remained awake and relatively alert and it was a pretty unpleasant experience. I had to lie on my left side whilst the colonoscope (a thin flexible fibre-optic telescope which allows the operator to see inside the colon) was passed through my anus, into my rectum and along my colon. Not long after the test began, however, I howled out in excruciating pain when I could feel the surgeon trying to push the colonoscope through my colon. Apparently I repeatedly asked the surgeon to stop the investigation.

When I came round from the sedation enough to converse coherently with my surgeon he explained that the reason the colonoscopy couldn't be successfully completed was because I had a *"redundant"* sigmoid colon. He explained that the sigmoid colon was loopy and long, grey in colour and had lost all of its muscle tone and elasticity. (In fact I even remember seeing my colon on the screen looking distinctly grey and not pink.) Further, he explained that the lack of muscle tone would have rendered that part of my colon functionally incapable of effective peristalsis. He described my sigmoid colon as being almost completely featureless.

Due to the inability to complete the colonoscopy I therefore had to have a further test and a *"virtual colonoscopy"* was performed that same afternoon. This was essentially an MRI scan of my colon, so that a full picture of my colon could be seen. For this test I had to lie on my side to have air pumped through a tube that was inserted into my rectum to make my colon bigger and easier to see, before lying on my back. This was incredibly painful but I knew I had to get the test finished so that I could get a clear diagnosis and receive some sort of treatment plan.

The virtual colonoscopy revealed that not only did I have a redundant sigmoid colon but also a redundant transverse colon. My surgeon then explained that whilst surgical intervention to remove the redundancies could be performed he wanted a second opinion on the best treatment plan for me. My surgeon also wanted me to see an endocrinologist to rule out any endocrine problems that could have been causing my night sweats. Until I had the second opinion and endocrinologist appointment he said that he wanted to try me with *"fleet"* enemas to keep the left side of my colon empty, which he explained would ease some of my symptoms.

For the first fleet enema I actually saw my surgeon in clinic and he administered it himself. I had to lie on a bed on my left side with my left knee bent and my right leg on top of my left with the bed tilted downwards so that my backside was slightly raised. My surgeon then inserted the tip of the enema bottle into my anus and rectum and emptied the 120ml of liquid into me. I had to then lie there until I could resist the urge to defecate no more. (Quite why my surgeon had to demonstrate how this was done I will never know because I'm sure the attending nurse would have been quite competent..... It did cross my mind that perhaps he enjoyed it – but I would rather not go there, particularly as he was old enough to be my father...!!!)

Anyway, once the urge to defecate became too strong to bear I went to the toilet and literally stayed there for the best part of an hour whilst my bowels emptied what felt like an enormous volume of fowl smelling faeces. It was painful and once I'd finished I felt weak and dizzy.

My surgeon then prescribed me fleet enemas twice a week and told me that I could use them more often if I felt like I needed to in order to keep my bowel as empty as possible and my abdomen as comfortable as possible. I found using the fleet enemas incredibly uncomfortable and the bowel movements that followed were often very painful and I would spend far too long on the toilet once I'd used them. Sometimes I would only be ten or fifteen minutes on the toilet but more often than not I would spend half an hour or an hour either on the toilet itself or to and fro once I'd used the enemas. It was certainly no way to spend my evenings once I'd put the girls to bed.

In accordance with my surgeon's instructions I saw a consultant endocrinologist, privately in mid-April. I had various blood and other tests and the endocrinologist ruled out any endocrine problems. This was a relief, but it also meant that my symptoms were almost all certainly related to the problems with my bowels.

It was the end of April before I was referred for a second surgical opinion and I was really struggling with my health by this point. What made it harder was that I couldn't ask my parents for support during this time as my father was critically ill with cancer and my mother was nursing him at home. And whilst my husband was incredibly supportive, he was running a new business which required his full attention. I therefore soldiered on as best I could, working as much as my health would allow and resting where possible.

On the up side, as I was working full time at this point (when well enough) my childcare arrangements were 8am until at least 8pm which thankfully meant that when I was too poorly to go into the office my girls were well looked after, either by my nanny, school or after school club.

I eventually had my first consultation with my new surgeon confirmed for just over three weeks after my virtual colonoscopy (during which time I was still managing to get to work but was increasingly struggling with my health).

Then the Sunday before this appointment my father died.

I think I managed to be strong through my own illness to this point because my father showed such amazing qualities through his terminal cancer. He was constantly upbeat and grateful to have lived such a wonderful life to the full. Indeed he was described in his obituary as bearing:

*"his terminal illness with serene courage, resolute stoicism and graceful acceptance".*

And whilst I was overcome with grief when he did pass I managed to spend some very special and quality time with my father during his last year. In fact, as my father was diagnosed with his terminal illness almost a year to the day he died, in many ways I felt that I mourned with him during the precious time we managed to spend together in those twelve months. Indeed, having spent this time with my father made his loss a little easier to bear. And whilst I still miss him every single day that passes I often feel his presence with or around me. I just wish he was here in person to continue to guide me......Instead I often think about how he would handle things and what he would do or say to me....

Anyway, when I attended my new surgeon's office for my initial consultation (just four days after my father's death) he took a full history, did the obligatory rectal examination and then told me that before any treatment plan could be arranged there were some other tests that could be carried out. He explained that on his rectal examination he could feel that the anterior wall of my rectum was slightly collapsed and it was for this reason that he required further information on the extent of the prolapse. One of the tests (and the most important test) he required was a defecating proctogram which would enable the radiologist to visualise the mechanics of my defecation.

I felt immediately at ease with my new surgeon but became a little frustrated after this first appointment because it then took another eight days before I had the defecating proctogram, during which time I was becoming increasingly ill. As a private patient I was rather hoping that the test would have been carried out more quickly, particularly as my health was continuing to deteriorate.

When the day of the test arrived, my desire to get diagnosed and appropriately treated was such that I didn't really feel embarrassed or intimidated by the nature of the test. The radiologist's assistant asked me to gown up and put paper pants on and then explained that what the test involved was for barium paste to be inserted into my rectum and my vagina and then for me to sit down on a *"make shift"* toilet and attempt to expel the barium paste from my rectum using the same technique I used when having a bowel movement.

So I did as I was instructed, lay down on my left side to have the paste inserted and then sat myself on the make shift toilet and *"pushed"* when asked do so by the male consultant radiologist. Pictures were then taken with an x-ray video which recorded the movement of my pelvic floor and rectum as I attempted to evacuate the paste. I could tell from the radiologist's reaction as he asked me to push one last time that my test was abnormal and when I asked him if it was normal before I left the room he said that it wasn't and that he'd send the results through to my surgeon that day so that I could discuss my care plan with him.

I didn't actually get to discuss the test with my surgeon for another six days, which felt like a lifetime as each day that elapsed I felt worse and worse. Then when I did meet him to discuss the results he confirmed that they were abnormal. My surgeon explained that the test showed that I had an internal rectal prolapse and also that the anterior wall of my rectum had completely collapsed. He also said that as this first test had revealed further abnormalities which were adding to my constipation there was no real need to perform any anorectal manometry at this stage because there was a clear indication for surgery.

My consultant explained that an internal rectal prolapse is caused by damage to the pelvic floor which can be a result of long term constipation and straining or can occur during/following labour and childbirth. As soon as he mentioned childbirth I instinctively knew that the prolapse occurred directly following my second daughter's birth. It was straight after her birth that I started getting increased wind and more sluggish bowels and more to the point I have also never felt properly well again since her birth. Further, it was only after my second labour that I started to experience frequent constipation; before that I had always been very regular, save for the odd occasion on holiday and during my pregnancies.

My consultant went on to explain that a prolapse can cause a backup of faeces in the colon which in turn will, over time, cause the colon to stretch and lose its muscle tone and that this is why I had redundancies in my colon. He also explained that I could have suffered some nerve damage either in labour or during previous surgery which could also add to the slow transit constipation.

My surgeon explained all treatment options to me. He explained biofeedback treatment as an option which is essentially the neuromuscular training of the abdominal, rectal, puborectalis and anal sphincter muscles to correct any incoordination of these muscles which results in difficult defecation. But because of my significant redundancies in the transverse and sigmoid colon his belief was that this would not be effective alone. Further, rectal irrigation using a home kit with warm water, although helping to relieve my symptoms, would not treat the underlying cause. A transanal placation would have helped to reduce my rectocele, but would not have dealt with my redundant sigmoid colon or slow transit constipation. The recommended treatment plan was therefore a resection rectopexy which involved resection of the redundant sigmoid loop to help with the slow transit of faeces through my colon and the rectopexy to help fix the rectum to the sacrum thus preventing further rectal wall prolapse and rectocele formation.

During this consultation it was also explained to me that there would be a primary anastomosis performed in theatre, being the joining back together of the two parts of the colon, in between which the redundant sigmoid loop used to sit and that the risk of leakage from this is approximately 5%. My surgeon also explained to me that if I were to have an anastomotic leak this carried with it a serious risk of morbidity and approximately a 2% risk of mortality. Additionally, if I were to have a leak it would likely require further surgery and formation of a colostomy, which I would have to live with for at least six months before a reversal could be considered.

Holy shit I thought to myself as he was relaying all of this to me ..... (no pun intended).

But despite these risks, I was completely happy to proceed with the surgery because I was becoming so poorly and just couldn't lead a normal life any more. I was also keen to have the surgery as quickly as possible because the pain and the nausea I was constantly suffering with were debilitating. And I figured that a 2% risk was sufficiently low for me to feel comfortable with the surgery because the reward of getting my life back on track was such a wonderful prospect. The way things were at that point I had no real quality of life anyway.

So after eighteen months of repeated hospitalisations, GP visits and visits to various consultants I was pleased that I was finally clear on why I was getting constipated and suffering with colicky abdominal cramping. I was, however, also bloody annoyed that the tests hadn't been carried out sooner. Why did I have to have five emergency admissions to the NHS with abdominal cramping and colicky pain and with x-rays showing faecal loading but without any further relevant testing carried out? And why did it have to take me going as a private patient to have the various tests done?

The sad fact of the matter is that during my NHS admissions all the doctors and nurses seemed to be bothered about was treating my symptoms and if the cause wasn't blindingly obvious (which of course mine wasn't) I was discharged as soon as I felt a bit better (which was usually after an enema).

I was fortunate enough to be able to afford private healthcare and was dogged in my determination to get a proper diagnosis. And whilst I'd like to hope that if I'd been unable to pursue the private route I would have eventually been diagnosed by the NHS I'm not so certain. And even if I eventually did succeed in obtaining a diagnosis it would no doubt have taken many months longer and I would almost certainly have had to have given up work pending any treatment as I was at the stage where I simply couldn't function.

Most alarming, however, was that I'd been repeatedly told by my GP and by numerous doctors I had encountered on their ward rounds along the way was that there was nothing wrong with me (other than a bit of constipation) and at times I was made to feel like a total fruitcake…….. Luckily my problems were mechanical in nature, but what if it had been cancer or some other deadly disease, pathology or infection? Would that have been missed too? If the constant bad press about neglect and misdiagnosis on the NHS is anything to go by then I suspect it would have been.

As I'm sat here writing, this week alone, there have been two articles in my daily national paper about neglect and failure to diagnose in time. One about a young graduate who diagnosed her ovarian cancer with an online search after months of visits to the doctor, who has only been given two years to live (but whose prognosis would have been excellent had she been listened to earlier) and another about a mother of four and young grandmother of eight (forty five years old to be precise) who died as a result of a treatable blood clot in her bowel because the nurses and the doctors on the ward where she was hospitalised were "*too busy*" to listen to her cries for help or her relevant history.

The family allowed a picture of her to be published in the press, bent over in the foetal position on the hospital floor in agony, begging for pain relieving drugs. The same article told of how this ladies' family had informed the medics of her history of blood clots (which can be diagnosed by a scan) but that a scan was not carried out until it was too late to save her.....

I digress....but the sad fact of the matter was that I had to face major surgery at only thirty eight years old due to the trauma I'd suffered in childbirth at the hands of the same institution that had so spectacularly failed to diagnose me. I felt both sad and angry, but was also keen to just get the surgery over and done with so I could get on with my life.

# Life changing surgery at 38

My operation was scheduled for the day after my thirty eighth birthday and the day before my operation (so my birthday itself) I could eat nothing at all and only drink clear fluids. I also had to take a dose of picolax which was as unpleasant as ever. Luckily, champagne counted as clear fluid …. So I could at least toast my birthday with a glass of bubbly. My day was obviously tinged with great sadness as I had only lost my father just over three weeks prior and I was nervous about my operation….

Fortunately the operation went really well and despite spending the first couple of days after surgery in the intensive care unit (**ICU**) at my local private hospital I made a good recovery, was discharged after six days and managed to return to work after only four weeks.

Unfortunately, however, my constipation symptoms were not completely resolved with this surgery. And whilst the severity of my symptoms had definitely improved I was still feeling nauseous, constipated and bloated and was still suffering with excessive wind.

In order to help deal with these symptoms my surgeon referred me to a specialist pelvic floor clinic run by a sister at my consultant's NHS teaching hospital. The sister gave me various pelvic floor exercises to strengthen my pelvic floor, but the main problem I was experiencing was continued constipation. As such the sister recommended that I start using the *"peristeeen anal irrigation system"* to keep my colon empty and provide relief from my symptoms.

Basically, what the system involved was pumping just over a litre of warm tap water up my backside whilst sat on a toilet and then waiting for my rectum and colon to empty. The indignity of having to use rectal irrigation twice a week at home to keep my colon empty was actually surpassed by the relief from symptoms. It was, however, a huge inconvenience for me, because it would take anything up to three hours from insertion of the water for my colon to empty. This meant that I could only really irrigate in the evenings after work or at the weekends once the children were in bed because from the moment I finished irrigating I had to be near a toilet.

After a couple of months of using the irrigation system not only was I becoming increasingly frustrated with having to use it I felt like it was gradually becoming more and more ineffective.

So I took it upon myself to do some on line research (using some decent medical literature) into what else could potentially help with my constipation. I discovered that sacral nerve stimulation (**SNS**) can be used as a possible treatment for constipation and because my symptoms were still so problematic I was willing to try it. What it involves is the insertion of an SNS wire through the sacral foramen in the back and the wire is then attached to a pulse generator which stimulates the sacral nerves. The stimulation of the sacral nerves should then increase the frequency of bowel movements. It is also used for both urinary and faecal incontinence and chronic pelvic pain and the reasons why it works for all of these problems are not fully understood.

What is known, however, is that the sacral nerves control the bladder, bowel and pelvic floor and the muscles related to their function. The SNS wire, once implanted, stimulates the sacral nerve with mild electrical pulses thereby enabling a person to perceive sensations that have otherwise been lost due to nerve damage.

But before a permanent device can be implanted into a subcutaneous gluteal pocket (upper buttock) it is necessary to undergo a temporary wire implant. For the temporary implant the wire is attached to an external pulse generator and if symptoms improve whilst the temporary implant is in place it is then viable to have a permanent device fitted.....

So in the October after my thirty eighth birthday I agreed with my consultant and the pelvic floor sister that I would undergo a procedure for the temporary device to be fitted. I had to keep a bowel and diet diary for the two weeks prior to the procedure and also for the two weeks following the procedure.

The operation to fit the device took place at the end of November. It was a relatively quick procedure and I experienced very little pain afterwards. Then during the two weeks whilst I had the temporary wire in situ I noticed a real improvement in my symptoms including the frequency and amount of my bowel movements. Whilst the improvement was not to the extent where my bowels returned to how they were prior to my second daughter, the improvement was enough for me to want to go ahead with the permanent implant (particularly because I hoped it would mean that I wouldn't need to continue using the irrigation system). I also experienced more sensation in my clitoris and less pelvic pain whilst the wire was in situ (which were unexpected but very welcome outcomes). I therefore scheduled to have the permanent device fitted as soon as possible for early in the New Year.

In terms of how or why the stimulation works for constipation, urinary and faecal incontinence, sexual dysfunction and chronic pain, the published literature is slightly hazy and as I said before it's not entirely understood. In my case, however, I'm absolutely certain it's because I suffered pudendal nerve damage when my second daughter became stuck at the level of my ischial spines and then had to be disimpacted from there for delivery.

The relevant anatomy is that the pudendal nerve is made up of branches from the anterior sacral nerve roots (S2, S3 and S4) and these nerves then join together to form a single nerve about 1 cm behind the ischial spine. As such, the stimulation of the sacral nerves has an effect on the pudendal nerve. The inferior branch of the pudendal nerve gives sensation in the rectum, the loss of which results in a lack of awareness of the need to defecate and thus results in constipation. Then the perineal branch of the pudendal nerve supplies the erectile and sensation functions in the clitoris. And as I said earlier I have suffered with a loss in clitoral sensation since the birth of my second child.

Now I'm not a medic, but I can understand the relevant anatomy enough to see that it's as clear as the light of day that my daughter's impaction damaged my pudendal nerve. This is one of the reasons why I find it absolutely astonishing that it would have been such an uphill struggle to bring a successful law suit for negligence against the hospital where I gave birth to my second child. My legal team was confident that there had been a breach of duty due to the length of time I was left in second stage labour and by leaving me so long my daughter had become impacted which in turn caused the damage...... But the sad fact of the matter is that the medical fraternity tend to close ranks and expert witnesses won't give claimant friendly statements unless a case is absolutely cut and dry or the damage so blindingly obvious that a favourable statement has to be given.

Anyway, back to my suffering.

Well, over the Christmas period in between having the temporary wire removed and the scheduled date for the permanent implant I had a couple of episodes of acute abdominal pain and nausea. Then on the first Thursday after returning to work in the New Year I began to feel increasingly nauseous and in pain. That night I took some tramadol for the pain and metoclopramide for the nausea, but I was in so much pain and so nauseous that I couldn't sleep. I therefore took myself to the A+E department of the NHS hospital where my consultant is posted and I was promptly admitted.

I stayed at the NHS hospital for two nights, during which time I was given IV fluids for dehydration, IV anti sickness for my nausea and IV morphine and other pain relief for the agonising pain. I also had a plain abdominal x-ray on admission (which is reproduced on the front cover) which showed faecal loading. I couldn't quite believe the x-ray though because I'd been using the anal irrigation system at least once or twice a week for the previous five months (with the exception of the two week period where I had the temporary nerve stimulation wire fitted). I was therefore beginning to fear that my colon just wasn't working at all.....

I was given various laxatives by the nurses including picolax to try and empty my colon completely but I didn't feel that they'd worked because despite some movement I still felt that I was loaded. Then after two days of no answers and no visit from my consultant (or indeed any consultant) I was really fed up of being on a ward in an NHS hospital. Granted it was the weekend, but I would have expected to see a consultant and his retinue on at least one of the days.

To make things worse I was on a twelve bedded general surgical ward and hadn't slept for two nights as it was so noisy. There was an elderly lady who cried out all night long and other patients who snored or called out for pain relief which made it virtually impossible to sleep. I therefore phoned up my private hospital to see if they had any beds and to see if my consultant would admit me under his care. As it was a Sunday the admissions sister said that I would have to pay a fifteen hundred pound bond before they would admit me which would then be refunded to me once my insurers authorised the care.

I really didn't care at this point whether the money was refunded or not as I just wanted to get out of the NHS hospital and into far nicer surroundings, with a duvet, my own room and en-suite and direct access to my surgeon.

Because I hadn't actually been discharged properly from the NHS and I was still on IV fluids I had to arrange for a private ambulance transfer before the ward sister would let me go....So I transferred with my IV fluids still up and shortly after my admission I had a CT scan.

When my consultant first visited me on the evening of my transfer the idea was to just keep me comfortable until the planned fitting of the permanent SNS device the following Thursday. However, once he'd seen the results of the CT scan he said that he wasn't sure whether the SNS device would adequately address my problems.

My consultant explained that the CT scan showed faecal loading in the ascending colon and that given that I'd already had picolax it was clear that the faeces couldn't move along the redundant segment of my transverse colon. As such, a more appropriate treatment would be the removal of my colon with the joining of my small bowel to my rectum, which is known as a sub-total colectomy with ileo-rectal anastomosis.

He explained that this was a major surgical procedure with risks of complications and death and that this would be an open procedure and as such I would therefore have a cut down the centre of my abdomen from my sternum to my pelvis. He also explained that I would suffer with diarrhoea after the surgery (for ever) and that I would have at least two to four episodes of diarrhoea a day (for ever).

Oh Joy.

During this consultation I sobbed uncontrollably because, whilst remote, I realised that there was a risk of death. And whilst this was also the case for my previous surgery under his care I knew that this surgery was a much bigger deal. I was also upset because I didn't want a "*bag*" if things didn't go to plan and that even if they did go to plan I would suffer diarrhoea for the rest of my life....

But by the day of my operation I was actually looking forward to going to theatre because I had been in hospital for six days already and was in constant pain and felt incredibly nauseous.

I'll never forget sitting in the waiting area before being taken through to theatre for that operation. I was feeling like shit and there was a very glamorous woman sat next me who looked as if she was wearing full make up (and who I assumed rightly or wrongly was about to go under the knife for plastic surgery) and she was trying to make light conversation with me. I really snapped at her when she attempted conversation because I was in so much pain and just wanted to get to theatre. I told her quite curtly that I was too poorly for polite pleasantries and asked her to please keep quiet! She gave me a startled look and promptly shut her trap.

Anyway, whatever my pain levels were when I was in that waiting room really paled into insignificance when I came around from my anaesthetic. When I woke up in recovery I can remember being told by my surgeon that the operation itself went well and thinking to myself really???? Because I was in such excruciating pain. I had been offered a spinal anaesthetic in addition to the general anaesthetic due to the extent of the surgery, but I declined because I didn't like the thought of being without feeling from the waist down and as I came around I was sorry I'd made that decision. But I was, at least, given patient controlled anaesthesia (**PCA**) which I could press at any time to give myself more IV morphine. I was then transferred from recovery to the ICU due to the risk of complications that the operation carried.

In the ICU I was continually monitored by an ECG, blood pressure gauge, temperature gauge, heart rate monitor and respiration monitor and I recall that on the first evening following my operation when my consultant visited he caught me straining to see the values on the monitor displaying my vital signs. He quipped that I shouldn't watch the monitor as I would scare myself, but he was also quick to reassure me that despite my low blood pressure and high heart rate I was doing extremely well and should:

"*try to relax and rest*".

I don't recall much about the three days I stayed in the ICU aside from the fact that even lifting my head enough to sip my first cup of sugary tea through a straw about twenty four hours after surgery was agony and that when the physiotherapist came to do some simple breathing exercises even that was excruciating. In fact the slightest movement of my body left me wanting to scream in pain…… I really didn't want the ICU nurses to bother giving me a bed bath but they insisted and it hurt like hell when they got me to move even the tiniest bit.

On the fourth day following my surgery I was moved to a high dependency bed just outside the ICU for one night. I barely slept that night because of the noise and the pain. My PCA had been taken down and I had to call the nursing staff for morphine, which didn't take long because I was on one to two nursing (rather than one to one as on ICU) but I wasn't getting the same immediate relief I did when in the ICU with my PCA in situ……. then the following day I was moved further down the ward into a quieter room.

My consultant was fully expecting me to only stay for two days on the ward before being discharged, particularly as he'd already experienced me recovering so quickly the previous May. Unfortunately, however, on post-operative day five I started to deteriorate. I began vomiting uncontrollably with real force and couldn't keep any food or drink down. My abdomen also started to swell up, making me look six months or so pregnant and my pain which had started to ease began to increase in intensity and my wound began to feel hot to touch. My temperature, which had previously been stable started to spike and as I deteriorated I became very scared because I felt even worse than I did when I first came around from the surgery. I'd also not passed any wind or stool since moving to the main ward.

That first night on the main ward (where I was on a higher ratio of patients to nurses) I barely slept and as I drifted in and out of a fitful sleep I started to hallucinate, both visually and aurally. I thought I could hear the nurses and doctors chattering outside my room and I could see the children in the seaside oil painting opposite my bed walk out of the picture. I'm still not clear why I was hallucinating but guess that it was probably just a combination of the morphine, antiemetics, infection and fever.

To make matters worse, my wound started to become unbearably painful overnight to the point where it felt like acid was being poured into it. I wondered whether the force of vomiting had actually ruptured some of my internal stitches….. And because I felt so unwell I constantly asked my nurse and nursing aid from the moment they started their shift to call my surgeon urgently. I knew I was deteriorating more and I was petrified. I knew I couldn't control what was going on in my body and even though I wasn't eating or drinking anything I was still retching and vomiting up even the smallest sips of water I was taking to keep my mouth moist.

When I eventually saw my consultant that evening in my ward bedroom I was so grateful that he was there but even in my poorly state I could instantly tell when he looked at me that he knew I was acutely unwell. He arranged for me to have an emergency CT scan explaining that he was concerned with my abdominal distension, fever and vomiting and that he needed to elucidate the cause. He also looked at my wound and said that as it was very warm and inflamed around the edges it must have become infected. He said I needed some antibiotics, which would have to be administered IV due to my vomiting.

I clearly remember my surgeon opening the stitches and taking a swab, which was pure agony, and I also remember the nurses saying that I'd have to be taken through to the x-ray suite immediately for my scan. I was writhing in agony from the pain in my wound and from my bloated abdomen, despite having just been given some oromorph, so I asked one of the nurses if she could run some IV paracetemol through my cannula as quickly as possible to try and relieve my pain. As I was wheeled around for my scan by the porter I had my paracetemol running, but was still in severe pain and still vomiting, so was holding a sick bowl as I went.

Just prior to the CT scan itself I had to have some dye injected into my arm and the radiologist explained that this was so he could ascertain whether the anastomosis site had leaked. Even being moved from my hospital bed onto the table of the CT scanner was absolute agony and it was difficult to lie still because my automatic reaction was to tense up because of the pain and move to get comfortable.

But I couldn't.

Thankfully I managed to stay still enough for the scan to be performed and miraculously I didn't actually vomit during the scan ….. but did so again immediately afterwards.

Once the scan was finished and the consultant radiologist had conferred with my surgeon, he told me that whilst they couldn't entirely rule out an anastomotic leak, it appeared that the anastomosis was oedematous (swollen) and I had a large collection of free fluid in my abdomen. The radiologist explained that I would have to have a drain inserted to aspirate the fluid and then some further scans at later dates to assess progress. I was in such pain, and was so hopeful that draining the fluid would help, that I just nodded when he explained that it involved my abdomen being cut under local anaesthetic with the guidance of ultrasound and a drain being placed in the incision. The drain would then enable the fluid in my abdomen to be aspirated and then subsequently drain out.

I felt barely conscious as the procedure took place and remember watching the radiologist in a daze as he cut me and collected some fluid. I then remember him clenching my right shoulder as I was leaving the x-ray suite and saying to me:

*"Now you hang in there won't you".*

When he said these words to me my consciousness suddenly kicked in because I realised that I was really very ill indeed.

Back in my room my surgeon told me that my small bowel was dilated all the way to the anastomosis and in order to relieve the pressure on the anastomosis they would need to empty as much of the remaining content from my abdomen as possible. To do this I would have to have nasogastric decompression. He explained that this involved a nasogastric tube being inserted up my nose and down the back of my throat to my stomach.

Having the tube inserted was horrific because I continued to vomit as the tube was being pushed up my nose by the Resident Medical Officer (**RMO**) and I was gagging as it passed down my oesophagus and into my stomach. I was also in acute pain at the time from the wound infection and the bloating in my abdomen and have had numerous traumatic flashbacks to this moment; partly because I was repeatedly retching during the insertion of the tube and so it was not an easy job for the RMO and nurse attending to me and partly due to the pain and discomfort I felt as it was inserted, coupled with the agonising pain I was feeling at my wound site.

The local anaesthetic from having the drain inserted was wearing off and due to the proximity of the drain to the wound infection the pain was intense. Also, despite the retching and gagging during the insertion of the nasogastric tube I was acutely aware that the tube had to be inserted or I would get even more seriously ill. As the tube was being pushed down and as the local anaesthetic was wearing off and my wound pain building all I could think of was Daniel Craig in the James Bond movie, *"Casino Royale"* where he is strapped naked to a seat-less chair and has his *"balls"* repeatedly whipped with a heavy knotted rope! I'd only just seen the movie for the first time a few weeks earlier and I must have subconsciously recalled it so I could comfort myself with the fact that the pain I was suffering from must have been less than his pain... but I also thought about my precious girls and husband who I had to be strong for.

I knew I had to get better and that I had no option but to stay still and grit my teeth and have the tube inserted. That night I didn't sleep at all despite being heavily sedated with morphine and having had a full strength sleeping tablet. I was scared that if I drifted off to sleep I wouldn't wake up. It was the strangest feeling.

I felt like I kept almost falling unconscious and I would then literally force myself to stay present and would almost jerk awake for a while. All the time I had aural hallucinations and when I started to feel more alive again later the following day I didn't actually realise that what I'd heard wasn't real because I started speaking to one of the health care assistants about what I'd heard the previous night and she just looked at me as if I was mad.

As I recovered more I also realised myself that what I'd heard that night was obviously all aural hallucinations. Several months after my discharge I had my cards read by a psychic medium and she made reference to me having been incredibly poorly and being pushed back by loved ones who'd already parted this world and being told by them that it wasn't my time. And despite my inherent scepticism about card reading and mediumship I truly felt that she was right. I had to fight for my life that night and I still get haunting flashbacks from time to time.

The nasogastric tube, at first, when aspirated by the nurses drained off nearly a litre of green and yellow bile. After that a smallish rectangular bag was placed at the end of the tube which just filled up with small amounts of bright green fluid with even brighter green flecks in. The nurses then aspirated the tube with a syringe each day when required. The bile, remnants of which were always around the top of the bag smelt like a rotting drain and made me feel even more nauseous than I did from simply being poorly.

It smelt so much that when my children first came to visit me one day during that long week when the tube was in situ, my eldest daughter ran straight over to the window holding her nose saying that she felt sick because of the smell. My husband promptly did a u-turn with our girls and took them off to M&S to buy two reed diffusers ……. One for my bedroom and one for my bathroom. The bedroom one dealt with the smell from the bag attached to my nasogastric tube and the bathroom diffuser helped to deal with the smell from my bile laced fluid diarrhoea. And once the diffusers had established themselves everyone who visited me from then on commented on how nice my room smelt!!!

That week when I had the nasogastric tube in situ and when I was at my most critically ill with the infection and the fluid collections, the doctors had increasing difficulty in cannulating me. And when a cannula was successfully inserted it would only last a day or so before it lost its' patency and finding a new vein was then becoming increasingly difficult. I needed to have the cannulas because I was on IV fluids (as I was nil by mouth), IV pain relief, but most important of all I was on two different types of IV antibiotics – metronidazole and tazacin, and I knew that if I didn't receive these I might die.

Having my veins start to shut down was incredibly scary as I knew this was a sign of how seriously ill I was. The only people in the hospital that could actually cannulate me during this time were the very experienced ICU doctors who looked like they were just stabbing my arm *"blind"* to reach better and deeper veins. These ICU doctors tended to be anaesthetists who were either retired from the NHS and wanted to keep their hand in, or they still held NHS posts, but wanted to work extra shifts privately to top up their pay. They were incredibly skilled as they managed to get to a decent vein each time.

Over the second and third postoperative weeks I also underwent repeated CT scans as my CRP wasn't going down quickly enough and I wasn't improving as quickly as my surgeon would have liked. I had a total of four drains inserted over this period into four separate fluid filled areas and whilst none of the drains seemed to be draining infected fluid (according to the microbiology tests carried out on the fluid) each time fluid was drained my pain would ease slightly.

Finally, when the fourth drain, which was inserted into a very tight space, but which was a site of intense pain, started to drain what looked like faecal matter, the radiologist thought that he had found an area where there may have been a leak from my bowel into my peritoneal cavity. However, after a few days had passed I noticed that this drain seemed to drain off fluid only when I had a bowel movement.

My normal consultant was actually on vacation when I realised this and I had one of his other surgical colleagues "*baby sit*" me in his absence. When I alerted my "*minder*" to this he ordered another ultrasound to investigate whether the drain tip had actually perforated my bowel (either during insertion or through migration after insertion). The consultant radiologist couldn't locate the tip on the scan and so he concluded the drain must have perforated the wall of my small bowel and so removed the drain. I was worried that with the removal of the drain my bowel wouldn't repair itself and that faecal matter would be escaping into my abdomen, but my "*minder*" assured me that the bowel ought to repair itself as the tip of the drain is so small. And whilst my CRP did initially rise again following the drain's removal it started to go down again quite quickly.

In terms of my bowel habit, I didn't have any bowel movements for about the first five or six days after the operation and then when my bowels started to work I was often too weak to make it to my en-suite bathroom and instead had to use a commode by the side of my bed. On at least two or three occasions I also woke myself up in the middle of the night having soiled the bed to find myself lying in smallish patches of greed liquid. Thankfully, being in a private hospital, my bedclothes were fairly promptly changed when I rang my call bell and I was too poorly to feel even the slightest bit of embarrassment about these instances of incontinence.

Once the nasogastric tube was taken out and I started to introduce soft and easy to digest food my bowel movements slowly turned from liquid green to liquid yellow and then brown, but even when my stools resumed their normal colour they were very liquid. As I started eating I was visited by the dieticians who spent time with me going through my new post colectomy diet.

The diet was low residue and low fibre to start out with and essentially consisted of white carbs and not brown. In addition I had a list of *"forbidden items"* including, no pith, no peel, no seeds, no skins, no leaves, no bubbles and no brown meat or pink fish. The dieticians also added in some energy milk shakes and smoothies to my diet because of my continuing nausea and general lack of appetite.

In fact food really wasn't that appealing to me during the early stages of my recovery and so being able to sip on a nutritionally complete and pleasant tasting drink was actually really useful.

When I became a little stronger and my CRP was very firmly on the way down my surgeon told me that my CRP had gone up to three hundred and ninety eight before I had the first CT scan and was started on the antibiotics. This is a seriously and life threateningly high CRP reading. A normal person has a CRP of less than five, but it can go up to about one hundred and fifty after surgery and around two hundred for serious burns and infections. People have been in the morgue having died of severe infections with lower CRP readings and it actually took several weeks for my CRP to drop to a more normal post-operative level. But even when my CRP was steadily dropping my recovery remained slow and I still felt incredibly weak and sometimes had difficulty even summoning up the energy to get out of bed to move to the bathroom.

One of the reasons I felt so weak was because my haemoglobin had dropped very low due to the severity of my post-operative illness. I hadn't actually had any internal bleeding; the anaemia was merely a result of having been critically ill, which, I understand, is an unexplained phenomenon that frequently occurs in ICU patients. So, before I could be discharged home I needed to have a transfusion of two units of plasma on two separate occasions.

I spent a total of thirty three nights in hospital during this period, two with the NHS and thirty one in my local private hospital. Because I was so ill during my hospitalisation the time actually went very quickly for me. But, unsurprisingly, it felt like a very long time for my husband, my mother and my children.

# Another slow recovery

When the time came to leave hospital I was collected by my niece who'd moved into our home during my inpatient stay to offer some stability to my girls and some help for my husband. Whilst I also had a very good and dedicated nanny who'd been with us since my youngest daughter was an infant, having my niece around felt more homely for my girls and she did the jobs my nanny didn't do, such as the supermarket shopping, cooking and washing. My husband was struggling with these jobs prior to my niece's arrival, having to juggle running his business with visiting me in hospital and keeping our girls as happy as possible.

I had to be wheeled out in a wheel chair by my niece to leave hospital because I was simply too weak to walk the entire length of the corridor to the elevator and then again from the elevator to the car. I was nonetheless delighted to be going home, albeit my delight was tinged with nervousness about how I would fare at home, particularly as I was still in a lot of pain and my wound still hadn't healed.

Because the wound infection was so deep, and the incisional site had to be opened up to let the infection out, my wound then had to heal from the inside out. Whilst in hospital my wound was dressed and packed by the ward nurses every other day and so on my return home I had to have home visits from my local district nurses to do this job. They continued to visit me to carry out my wound care at home until I was well enough to drive myself to the GP surgery for my care to be continued there.

Initially on my return home I slept for most of the day and night....I think I slept an average of around sixteen hours a day that first week back at home.

I was just physically and emotionally drained.

I couldn't walk very far because I was so weak from the surgical trauma, the severe post-operative infection and the muscle wasting in my legs from being bed bound. In fact my mum had to bring my dad's old wheelchair up to the house for me to use to get fresh air at the weekend with my family until my strength returned.

After about two weeks I was strong enough to start walking further distances and had no need to keep using the wheelchair to get out and about but at around the same time I started getting anxious and having flash backs to the night I thought I might die. And despite my exhaustion I started finding it increasingly difficult to get restful sleep, often waking with a start.

I therefore booked an appointment with the consultant psychiatrist who I'd been seen by on several previous occasions over the preceding ten years for intermittent anxiety. In fact prior to my surgery I was taking venlafaxine for anxiety but when I explained my recent surgery to my psychiatrist he explained to me that as these tablets are slow release I may not have been absorbing them in my small intestine and as such they may not have been working at all. He therefore changed my medication to duloxetine which is absorbed far more readily. He also referred me to have some counselling because of the extent of the trauma I'd just suffered.

The counselling was therapeutic to a point but I only went three or four times because I felt that once I'd talked over the trauma of my recent physical health there wasn't much else I wanted to discuss, plus I was starting to feel the benefits of the duloxetine.

As time progressed my recovery from the surgery quickened, although my wound had not healed well as it had "*over granulated*" and to be honest, it was a pretty unsightly scar. I was also feeling constantly thirsty, dizzy (particularly on standing from sitting), and nauseous.

My mum said that that my thirst, light-headedness and nausea were probably symptoms of dehydration due to the amount of diarrhoea I was having; I was opening my bowels around twelve to fifteen times a day and it was really watery diarrhoea. So I went back to see my consultant who said I could take up to five sachets of dioralyte a day but that I should also try some loperamide to slow my bowels down and firm up my stool. My consultant said I needed to aim to get down to around four to six bowel movements a day.

Taking the dioralyte certainly seemed to help and after about five weeks at home I was starting to think about my return to work. I'd been discussing my return with my line manager and Occupational Health (**OH**) and I set a date for my return which was almost three months to the day of my colectomy.

But just under three weeks before my return I noticed that I'd been having much less diarrhoea than usual and I started to get excruciating back ache, pain in my rectum and acute abdominal cramping. Then my diarrhoea stopped altogether, my abdomen became very distended and I started vomiting.

I therefore telephoned my consultant's secretary who advised me to go to the ECC at my local private hospital, following which she said I could be admitted, if required, under my consultant's care via the ECC.

At the ECC I was given IV morphine and IV antiemetics and then sent for an x-ray which showed that I had dilated small bowel loops throughout my small bowel, consistent with a small bowel obstruction. So I was admitted and was treated with IV fluids and nil by mouth. Fortunately, the blockage cleared quite quickly once the fluids were up and running and I was discharged home after only a few days. I was, however, very against taking the loperamide after this episode because my logic was that I would prefer diarrhoea every day then run the risk of another obstruction.

My surgeon tried to reassure me about the loperamide, explaining that he thought this would be a one off and that it was just everything settling down after the surgery and that I shouldn't be worried. I therefore kept taking the loperamide, but on a lower dose and returned to work as legal counsel at a major plc on my scheduled return date, initially working two days a week in the office on shortened hours with a view to upping my days and hours as I became stronger. The support I received from my employer via their OH lead nurse was absolutely fantastic and I felt very fortunate to be in a position where such support was available.

I relished returning to the office because I could concentrate on something other than simply getting better and after a couple of weeks back at work I started doing some gentle exercise at my gym. I was really pleased with how my progress was going and felt that I was getting stronger by the day, but about nine weeks after my last discharge, in early June, I started having sharp abdominal cramps again one Saturday evening after we'd had some friends over for a BBQ.

At first I thought the cramping was because I'd just eaten too much, but by one a.m. on the Sunday morning the pain had become so bad that I called a taxi to take me to the A+E department at the NHS hospital where my consultant works. I was promptly admitted after the initial x-ray in A+E revealed that I had a sub-acute small bowel obstruction. I was given IV fluids, but no one bothered to tell me to stay nil by mouth and so by the Sunday evening my abdomen had become very distended from only a small amount of fluid and a couple of pieces of toast and a repeat x-ray showed that the sub-acute obstruction had progressed to an acute small bowel obstruction.

I was very frustrated on the NHS ward where I was admitted. It was an emergency surgical admissions ward; a type of *"holding bay"* for patients admitted via A+E, who needed hospital care, but who had not yet been allocated a bed on the appropriate ward. The nurses therefore had to deal with a high turnover of patients with very varied ailments and didn't really have the time to deal with the number of patients under their care. Further, there appeared to be no resident doctors on the ward and instead, the doctors with the relevant specialism would be called to see to patients in between their *"on call"* work. With it being the weekend, I had to wait for what seemed like forever for anything to happen.

Time did, however, go relatively quickly, largely because we had a complete character in one of the beds in the small four bedded holding bay to which I'd been admitted. On my left was a young Italian student who'd been on two courses of oral antibiotics for a kidney infection which hadn't resolved, and so was admitted for IV antibiotics with known sensitivity to her infection; opposite was an older lady who had a dental abscess under an unerupted wisdom tooth which was so large it was almost obstructing her airway; and then opposite was the young (late teenage) girl who was admitted for a washout and repair of a deep inner thigh wound which she'd sustained climbing over a spiked railing fence whilst *"running away"* from a guy she'd spent the night with after meeting him for the first time at a house party!

And despite all of us feeling poorly she had the three of us in stitches. Above her bed on the rectangular white board which is used to display the patient name, allocated nurse and allocated surgeon, she'd written in the surgeon spot:

*"any young, good looking male doc will do ☺"*.

And the recounting of her adventure was truly hilarious, despite her injury! She explained to us that she'd left her previous night's conquest's house via the back door about five a.m. wanting to get home. The back door faced playing fields, the other side of which was a main road where she figured she could get a taxi. So on leaving the house she used the playing fields as a short cut to the main road, but when climbing over the metal spiked railings adjacent to the main road she speared her thigh.

She described to us in detail how she was stood on a ridge on the railings, running about a foot below the top of the spikes facing the main road, about to climb over the top of the railings. But when she attempted to bring her second leg over the top of the railings she lost her balance and a spike from one of the railings literally speared her thigh, like something from a *"horror movie"*.

After the initial shock, where she also lost her balance she said she could feel very little but found herself upside down, supporting herself with her arms outstretched holding onto the railings near the bottom, her skirt covering her face, one leg on the ridge on the fence trying to support herself, and the other ensnared on the railings with blood gushing out. She said for some unknown reason, apart from the initial impact, it didn't even hurt – she said her leg just felt completely numb and she only felt pain on moving, which she knew she shouldn't do……. A male passer by then stopped to offer his help and once he realised that she was impaled on the railings he called the police, an ambulance and the fire brigade.

The passer by stayed chatting with her until the fire, police and ambulance services arrived and she then explained that in order to be released from the railings a fireman had to use some sort of chain saw to saw through the metal fence, whilst being careful to leave the length of the spike that had speared her leg and a bit extra in her thigh (explaining that it would need to be removed later on at hospital).

She explained that whilst the fireman was sawing through the fence she had to be supported by a police officer and in order to get the best angle to saw through the fence she essentially had to sit on the police officer's lap….And as the fireman was sawing through the fence the vibrations were passing through the girls' body which was then involuntarily (?!) gyrating on the policeman's lap!

Needless to say, after several minutes of gyrating the police officer could no longer prevent nature taking its course and he developed a *"whopping hard on"* (to repeat my patient neighbour's exact words)! Before she was then taken to hospital the police officer asked for the girls' phone number and he later tracked her down and came to visit her on the ward …..!

But aside from the light relief provided by my neighbour I was thoroughly unhappy on the ward, particularly as by the Sunday evening I hadn't even been seen by a senior doctor. I was also really concerned that this was my second small bowel obstruction since my discharge that February and the standard of nursing on the ward was not good.

The young girl with the railing injury was on eight hourly IV antibiotics to deal with the infection that had started in her wound, but when her six o' clock dose was due her cannula had become blocked and had to be removed. But rather than immediately re-cannulate and give the antibiotics as required the nurse simply told her that as her cannula was no longer patent she'd have to wait until the morning to be re-cannulated and could start back on her antibiotics then…..!

Clearly the nurse either couldn't be bothered to cannulate her or was incapable of doing it and neither could she find someone else to do the job. And when I mentioned to the girl that she really ought to insist on getting the cannula re-sited and for her antibiotics to be administered she just said that she felt OK and that she was sure she'd be fine...

Meanwhile the lady opposite me had been complaining to the nurses all day of fever and pain and said she felt like her abscess was getting bigger but had been repeatedly told not to worry. Then she eventually insisted on seeing a doctor when her throat felt like it was swelling. When the doctor arrived he commented that her abscess was so large that she was in danger of having her airways completely obstructed and he was visibly cross with the nursing staff for not alerting him earlier, immediately took a swab and started her on a wide spectrum IV antibiotic until he got the cultures back…..

Against this backdrop of neglect (which was either due to poor nursing or poor staffing levels) I called my local private hospital who then contacted my surgeon to see if he would admit me again under his care.

As before, the NHS hospital wouldn't allow me to be discharged unless I signed myself out and was transferred by a private ambulance. They said that I wasn't down for discharge and needed my IV fluids and so would have to be transferred with them up..... But once the transfer had been organised I was collected fairly promptly by ambulance and taken to a room, which to be honest, felt a bit lonely after the company I'd had on the NHS ward. But on balance I was happier because I knew that my care would at least be very firmly consultant led as I was now in the private sector......

I was obviously hoping that the obstruction would resolve quickly again and that I'd only be in hospital for a couple of days, but unlike the first time this obstruction was proving more difficult to resolve. And after three and half days on IV fluids and nil by mouth I was still feeling obstructed, so my surgeon ordered an MRI scan of my small bowel.

When my surgeon came in to see me the morning after my scan he sat down beside my bed, but before he opened his mouth to tell me the *"normal"* results of the scan I just came out with:

*"it's an anastomotic stricture isn't it?"*

I really had no idea at the time where this came from.

Now I know this sounds completely mad, but I absolutely believe that my dad must have been watching over me and talking to me in my sleep because I didn't even know what an anastomotic stricture was, even as the words were leaving my mouth.

When I said this out loud my surgeon looked at me puzzled and said:

*"No"*

in a rather odd way in which he really lengthened the sound of the "o". But then he quickly added that the scan revealed:

*"no significant stricture".*

He had the scan report with him and read out load as follows:

*"the anastomosis between the ileum and the rectum is unremarkable with no significant stricture"*

He then added that the concluding remarks were:

*"Normal small bowel study. No significant findings. No obstruction or collection seen."*

He then went on to say, however, that if the scan missed something and the stricture was actually causing the obstructions it would be very easy to carry out a flexible sigmoidoscopy to look at the anastomosis. If he then found that the anastomosis had shrunk and was causing the obstructions he explained that it is easily fixed by balloon dilatation.

He explained that for this procedure a scope is inserted into the rectum and then a balloon is inflated with water at the site of the stricture. The inflation of the balloon at the stricture site then excerpts equal amounts of pressure around the circumference of the stricture causing the scar tissue to break down and the anastomosis to stretch. He explained that if he does find a stricture at the anastomosis site then this procedure would need to be repeated several times to give lasting results.

During this visit I also explained to him that the site of my earlier wound infection was giving me significant pain. My surgeon advised me that due to the collections in my abdomen after my operation there were probably significant adhesions under the surface scar which may need to be separated at some point in the future to give relief from the pain. He did, however, say that it was not appropriate to be carrying out such a procedure so early in the post-operative period. In addition, my surgeon said that there was also a chance that there was still some residual infection because I was spiking fevers and so he put me on a two week course of antibiotics to address any lingering infection.

I also asked my surgeon outright if he would take a look at the scan himself to see if he could see anything on there before taking me down for a flexible sigmoidoscopy. I explained that I was terrified of anything going near the anastomosis because I didn't want it to be disturbed and rupture. My biggest fear (aside from death of course) was of ending up with a *"bag"*. I was still only thirty nine and had a relatively good bikini body (at that time) and enjoyed keeping fit and being able to wear figure hugging clothes that could flatter my figure. I also worried about intimacy with my husband if I ended up with a bag and really just didn't want to have to deal with the prospect......

So my surgeon then went away and looked at the scan and when he came back that evening he said that he believed that the stricture was significant enough to be causing intermittent obstructions. He said that on his measurements on the scan it looked to him that the anastomosis had shrunk down to less than a centimetre. I asked him why the radiologist hadn't picked this up and he just shrugged his shoulders.

(One thing I have found on my journey is that doctors are very wary of criticising their peers.)

So I said that it wasn't acceptable for a radiologist to report that something was insignificant if it is clearly giving me symptoms....

My surgeon remained quiet on this point and just went on to explain that when the anastomosis was originally formed it was two point nine centimetres in diameter, but as he now reckoned it was less than a centimetre he felt that this would be contributing to my symptoms. My surgeon also explained that anastomotic strictures most often occur where there has been a leak at the time of the operation; so although no evidence of a leak was found on the CT scan when I had the initial post-operative problems, the very fact that I had a stricture seemed to indicate that a leak may have occurred shortly after the operation.

During this visit from my surgeon I was really upset and very scared.

I told him that I was worried that there would be a problem with stretching the stricture and that I would end up with a bag. My surgeon tried to cheer me up by saying that everything would be fine and that the only bag I would end up with would be a *"Louis Vuitton"* or *"Prada"* handbag from my husband!

But I was also starting to have a crisis in confidence with my surgeon at this point. Even though I had built up a good rapport with him and he made me feel that I was in good hands (because he was a good listener and always calm and confident) I was struggling to believe that all my post-operative complications were just bad luck.

So the night before the first dilatation I was crying inconsolably and reached out to my father's favourite and most skilled old senior registrar and I talked through my fears with him. He had become a respected and eminent consultant in his own right and had actually recently retired. But the telephone call was so worth making because my fears were dealt with and he actually gave me a renewed confidence with my surgeon.

One of the senior nurses on the ward also suggested that I might like a glass of wine that night to calm my nerves and to help relax me for my procedure the following day. So she asked the sister in charge for her consent to *"prescribing"* the wine and thankfully she agreed.

And I have to say that that glass of dry white and the call with the old senior registrar of my father's was the best combined medicine I'd had that day!

Then when it came to the morning of the endoscopic balloon dilatation I was feeling much more confident that everything would go well and I asked for as little sedation as possible as I wanted to see what was happening on the monitor. I lay on my left side and could see the scope go in on the monitor but then I remember nothing else apart from being in severe pain when the balloon was inflated.

After the procedure my surgeon said that the stricture measured 0.9cm at the start of the procedure and that that he'd managed to stretch it to 1.8cm in diameter. He told me that he wanted to get it up to as near to 3cm as possible and that I would likely need a further two to three dilatations. The next dilatation was scheduled for two weeks' time and a further dilatation was then booked in for approximately two to three weeks after that.

But by the time of the second procedure the stricture had narrowed again to approximately 1.2cm. So the stricture was stretched to 1.8cm once more and as I was still feeling rather dehydrated despite drinking regular dioralyte my surgeon arranged for me to have a litre of normal saline IV during this admission and during each subsequent admission for the balloon dilatations.

By the time of the third procedure, three weeks later, the diameter of the stricture had remained at 1.8cm which was great news. But during this procedure the stricture could only be stretched a further 0.15cm to 1.95cm because the pain was too intense for me to endure any further stretching, despite the fact I'd had IV sedation.

Then immediately after my third endoscopic balloon dilation the frequency of my bowel movements increased and by the following week I was feeling so dehydrated that I had a metallic taste in my mouth, was feeling dizzy (especially on standing), nauseous and had severe pain in my left loin. I was already drinking up to five sachets of dioralyte a day and so I couldn't understand why I still felt so weak with dehydration.

I tried to get in touch with my surgeon via my local private hospital but was unable to do so and so was rushed to my nearest NHS teaching hospital A+E where I was admitted and treated with IV fluids and pain relief.

During this stay I was referred to a consultant gastroenterologist who came to see me together with a dietician and they put me on a more structured fluid replacement regime of five sachets of dioralyte a day (rather than up to five) plus a fluid restriction of only one and a half litres of other fluids together with 20mg omeprazole twice a day.

The gastroenterologist explained that for optimum effectiveness some people needed to drink the dioralyte double strength as this would give a sodium concentration of 120mmols/litre which would make for easier absorption, but he felt that at the moment I would be OK with just the single strength provided that I was careful with the fluid restriction.

The dietician explained that I needed to avoid caffeine and alcohol (as much as possible) because they were stimulants and the reason for the fluid restriction was to guard against *"washing"* the electrolytes out of my body. The dietician also reinforced the low fibre, low residue diet to guard against any further obstructive episodes.

The gastroenterologist explained that the omeprazole was prescribed to cut down the amount of acid produced by my stomach which would, in theory, reduce the amount of liquid in my bowel movements. And after increasing the dioralyte to five sachets every day (instead of *"up to"* five), limiting my other fluids and taking the omeprazole I did certainly start to feel a little better in terms of hydration.

I did, however, feel a bit let down that my surgeon hadn't referred me to a gastroenterologist himself or to a dietician again after my recent obstructive episodes and so after I was discharged from the NHS hospital I asked him to organise both of these appointments for me. It made sense for me to be seen by a gastroenterologist and a dietician that my surgeon knew and worked with because this would give me *"joined up care"* which would hopefully help. In particular, I was becoming rapidly aware that I was a complex case and wanted to have a team in place that knew each other and worked well together.

So, when I saw the gastroenterologist who worked closely with my surgeon he wanted to exclude any other reasons for my diarrhoea. The reason for this was because compared to other patients with an ileorectal anastomosis, my frequency of diarrhoea was higher and my propensity to dehydrate was more marked.

Initially he simply organised tests for lactose and glucose intolerance. But as both tests were negative the gastroenterologist drew the conclusion that my diarrhoea was purely a result of my anatomy and would be something that I would have to learn to with. He advised me to try the antidiarrheal medication again, but I didn't want to take it because it hurt at the site of my anastomosis when my stool thickened up and I also got cramping pain elsewhere in my abdomen. So I made a conscious choice to live with my diarrhoea as opposed to being in pain and having to take analgesics every day.

The dietician I saw gave me useful information sheets about eating the right foods for intestinal strictures and also told me to start drinking the dioralyte *"double strength"*, so instead of five sachets in a litre of water, I needed to drink five sachets in half a litre every day. She explained that I needed to be managed like a high output stoma patient due to the amount of diarrhoea I was having and that the double strength dioralyte would give a sodium concentration of 120mmol/L which would be more readily absorbed.

The dietician also advised my surgeon to check my urine sodium to monitor for sodium depletion (as this is a more accurate test than serum sodium tests) and that if my urine sodium fell below 20mmol/L the double strength dioralyte could be increased to 1000ml/day (i.e. ten sachets of dioralyte) until my urine sodium rose to above 20mmol/L. She didn't advise on a particular fluid restriction but encouraged me to reduce the amount of other fluid I drank and to add salt to my food.

After this appointment I had my urine sodium checked at my GP but because the result came back at 40mmol/L (which although low was not below the lower end of the threshold) I didn't actually have it checked again that summer. (More on that later....)

In terms of my stricture, my last balloon dilatation procedure was scheduled for after I returned from my summer vacation, which I took in England for the first time in living memory. I had a *"staycation"* because I was petrified about travelling abroad and getting another obstruction whilst away and then being treated by medics unfamiliar with my history and where there may have been a language barrier.... Thankfully, however, my vacation was uneventful in terms of my health and I managed to have a fantastic break.

On my return, and during my fourth endoscopic balloon dilatation, when my surgeon inflated a 2cm balloon, the inner surface of my small bowel at the site of the stricture didn't even touch the inflated balloon and my surgeon estimated the new diameter of the anastomosis to be around 2.3cm. So, whilst this is narrower than the original 2.9cm diameter of the anastomosis created at operation, it wasn't too much smaller.

The relief I felt on hearing this news following the procedure was immense and I finally felt that the end was in sight in terms of procedures and hospital stays.

Unfortunately, however, I still had some way to go.....

# Back to theatre

I managed to stay relatively well for the next six weeks, until the middle of September following my thirty ninth birthday.

The first signs of going downhill again were that I'd noticed my bowel movements becoming a bit more watery and I also started feeling nauseous and in pain after eating which was also triggering a "*hot flush*" with sweats.

My night sweats also returned and by the last week of September I'd developed an unquenchable thirst and had deep pelvic pain and back ache. I therefore contacted my surgeon and asked if he could see me as soon as possible.

I knew that some of my symptoms were almost certainly due to dehydration but I was also concerned about the pain after eating and wanted things to settle as quickly as possible, particularly because I was starting to take on more responsibility again in work. The last thing I needed at that point was to be hospitalised again for a long period of time which would invariably interfere with my career once more……

When I saw my surgeon that Sunday afternoon he said that I was probably just dehydrated and he was hopeful that I could be treated with some IV fluids overnight and would be able to go to work the next afternoon. He explained that, for some unknown reason, pain often felt more acute with dehydration and that once I was back to euvolemia I should feel much better. Unfortunately, however, by the time my surgeon returned to see me on the Monday morning planning to discharge me my pain had significantly worsened and when he read through my blood results he saw that I was suffering from low potassium, which would also need treating before he could discharge me.

I therefore had some more fluids, which were infused with potassium, but rather than feeling stronger as the day went on I became progressively weaker, couldn't tolerate any food and had started passing bright green stools. Additionally I had started to suffer with a very colicky cramping pain in my abdomen with an almost constant and excruciating deep pelvic pain which was radiating to my lower back.

I was therefore kept in hospital for further tests to see if a cause for the nausea and pain could be elucidated. The tests I underwent were a plain abdominal x-ray, an MRI scan, CT scan, barium follow through (which involved the insertion of a naso-gastric tube again) and a further endoscopy to check my anastomosis. What the tests revealed were that I had what looked like an internal hernia (which was almost certainly causing the pain after eating and nausea), a cyst in my remaining left ovary and one in my left fallopian tube and that I also had adhesions in my upper abdomen.

My surgeon discussed with me the risks of more surgery, particularly the risk of developing more adhesions, but because the internal hernia was making it difficult for me to eat and because there was an inherent risk of this hernia becoming strangulated or causing a complete small bowel obstruction I made an informed decision that the risks of the surgery were worth taking if it meant that I could start eating again and leave hospital.

So just less than nine months after my colectomy I underwent a further major surgery.

Once again the surgery was open (rather than by laparoscopy) and at surgery various bands of adhesions were divided, my left ovary, fallopian tube and some of my omentum removed, a mesenteric window closed (in which small bowel had been getting intermittently stuck, causing the obstructive symptoms) and the site of the old wound infection which had left me with a particularly ugly scar was refashioned.

Thankfully I made a quick recovery from the surgery and was discharged five days later, although in total I'd spent a further twenty one days in hospital. My surgeon was, however, now very hopeful that following this surgery everything should settle down and that there would be a good chance that my stools would start to thicken up and I would be less prone to dehydration.

Unfortunately, however, only a few days after discharge I developed a visible haematoma under my mid line incision. My youngest daughter had actually given me a really enthusiastic hug shortly after I returned home and had knocked my incision with her knee which may have caused the haematoma…… but the area also became increasingly warm to touch and I started to get a fever and increasing pain and nausea.

So less than a week after my discharge I returned to the ECC where I was given pain relief and antiemetics and sent for an ultrasound scan. The ECC doctor felt that I might have had a collection of fluid in my pouch of douglas that had become infected and he wanted to get a scan so that this could be ruled in or out.

So prior to my admission onto a ward I was sent up to the imaging department for an abdominal and pelvic ultrasound. The radiologist on duty was the same radiologist who had said that my stricture was *"not significant"* earlier that year, when it had in fact shrunk to less than a third of its original size, so I immediately had no confidence in him.

I therefore asked very politely if he could talk me through the scan as he performed it. So he covered my abdomen in jelly and turned the screen towards me as he started to move the transducer probe across my abdomen.

After a couple of seconds when the transducer probe was over my haematoma there was a black area apparent on the screen and he said to me:

*"that is your uterus"*

and he then moved the probe across to the right a little and said:

*"and that is your right ovary"*.

I couldn't actually believe what I was hearing ..... I'd had my uterus removed five years earlier and my right ovary removed three years after that and so I quickly and curtly remarked:

*"really? I actually have neither! Are you sure the black area isn't fluid in my pouch of douglas? And that's why I'm in so much pain?"*

I then started to cry and said that I didn't want him to carry on with the scan because I had no confidence in him. I told him that on an MRI scan earlier in the year he had said that I had no significant stricture, but that the stricture was less than a centimetre and was causing recurrent obstructions and that after what he'd just said my confidence was shattered.

He just said to me:

*"ultrasound can sometimes be difficult to interpret"*

to which I said:

*"then why the hell is it used as a diagnostic tool?"*

I gathered myself together and the healthcare aide who was assisting the consultant radiologist was really lovely. She'd been a nursing aide on the ward when I was really poorly earlier in the year after my colectomy and we'd really bonded. She could see that I was weak, in pain and really frustrated. I apologised to her for being cross but she just put her arm around me and said that there was nothing to apologise for and that she completely understood.

When I got back to the ECC and told the doctor what had happened he said that he would speak to my surgeon and rearrange for another scan by a different radiologist the next day. I also told him that I was feeling really dehydrated and in pain and so he organised some IV fluids and regular pain relief before transferring me to a room on the ward.

When I got to my room I did some internet research on the consultant radiologist who'd seen my *"phantom"* organs and was alarmed to find that he was also a consultant at the leading teaching hospital in my area and the lead for cancer and gynaecological imaging and also an approved trainer for *"Gynae-Radiology"*......??!! What the hell??!! If he was the best in the area I dreaded to think what the others would've seen (or not) on my scan.

And when I saw my surgeon that night in my room he seemed almost embarrassed about what'd happened when I explained the episode to him and he didn't say much – although I thought to myself *"what could he say?!"*. He simply arranged for a repeat ultrasound the following day, which confirmed the fluid in my Pouch of Douglas, but which, my surgeon later explained could just have been the Adept® (an adhesion reduction solution) which they left in my abdominal cavity at the end of the surgery so as to reduce the chances of any new adhesion formation.

I was, however, given some antibiotics in case there was any infection in the fluid or my wound (given the warmth of my wound site) and after a couple of days on IV fluids, pain relief and antibiotics I felt well enough to return home again.

But within a couple of weeks of this discharge I fell poorly again.... ARGHHHH. I started to feel increasingly nauseous and dehydrated and was unable to pass any stool, had deep pelvic pain and low back ache and generally felt exactly the same as I had done prior to each earlier episode of obstruction (including the one directly before my last operation). I was completely despairing and was wondering if I would ever be well.

My surgeon was out of the country at this time and so I couldn't get a quick admission under his care. So I went to the A&E at his NHS teaching hospital where, one again, I was diagnosed with a sub-acute small bowel obstruction. I was given IV fluids, pain relief and antiemetics and after three days I had started opening my bowels.

On my surgeon's return to the country he was keen for me to stay on the NHS ward so that I could be reviewed by his Professor colleague and the head of his surgical department before considering any transfer to my usual private hospital. I therefore stayed on the NHS ward and was reviewed by the Professor who said that he was sure that this episode was:

*"just everything settling down after all the trauma to my bowels"*

and that the episode would pass. He also said that he couldn't rule out this happening again and that I should try and manage things at home as best I could and that when I couldn't I should simply come back in again.

He also suggested that I take up some yoga or pilates and practice meditation, which angered me slightly because I felt that he was trying to tell me that some of my physical pain was in my head. And whilst his general prognosis in terms of having to expect repeat episodes until things calmed down was not music to my ears, he at least gave me a realistic expectation of what I might have to face over the coming months. He also said that as soon as my pain had eased and the nausea had settled enough for me to eat and drink properly he was happy for me to go home.

However, as my nausea wouldn't settle and my pain was still quite acute I agreed with my usual surgeon that I should be transferred across to my local private hospital under his care again. At least I knew the staff well there by now, would be more comfortable (in my own private room with a duvet etc) and would get some sleep.

But after three days back there I was still feeling too poorly to go home. My pain had eased somewhat but the nausea hadn't and it would only settle enough for me to eat if I was given IV antiemetics an hour or so before food. I was also still feeling dehydrated despite having been given at least six litres of IV fluids in the previous week. And because it was becoming apparent to my surgeon that my problems were medical rather than surgical he referred me back to the gastroenterologist I'd seen that summer.

The gastroenterologist carried out some further blood tests and performed a gastroscopy in order to rule out any problems further up my gastrointestinal tract. The further blood tests and gastroscopy didn't reveal any new pathology and the continuing diarrhoea and nausea were put down to the loss of my colon and *"possibly"* to dehydration. I was given no new advice about how to deal with the nausea and diarrhoea and so I accepted that this was just the way things were going to be for me from here on in……….

At this time, neither my surgeon nor my gastroenterologist monitored my urine sodium or suggested that I increase my dioralyte intake to ten sachets a day, which with hindsight, they should have done on the back of the advice from my dietician in the summer. And I didn't even think to go through all my letters again containing the various advice I'd received and take it upon myself to increase my dioralyte….. I trusted the consultants' who cared for me and thought that if I needed to increase my dioralyte they would have recommended it…..So I simply continued with the half litre of double strength dioralyte a day and limited my other fluids to around a litre and a half.

# Feeling dry

Following my last discharge in mid-November I managed to keep my nausea under control with antiemetics and my dehydration at bay (or so I thought) with double strength dioralyte and minimising my other (hypotonic) fluids.

And on this regime I managed to have a great Christmas and New Year at home with my friends and family. I was able to attend most of the gatherings we were invited to and to keep up with my girls' busy schedule. I even managed to make it to the gym for some gentle exercise. On Christmas morning we took our girls swimming at our country club followed by Christmas dinner at home with just the four of us which was really special, particularly as I'd spent over four months in hospital throughout the year.

I was due back in work on the Tuesday following New Year and was actually really looking forward to getting stuck in, having been absent for so much of the previous year. But when I woke up on the last Sunday of my festive break I was incredibly nauseous and dehydrated. I'd had drenching sweats throughout the night and when I stood up to get out of bed I almost collapsed.

I'd also noticed a slowing of my bowel movements in the previous twenty four hours or so and had started to experience excruciating abdominal cramps. By mid-morning I felt so weak that I told my husband I needed to go into hospital. His initial reaction was to tell me that I couldn't just run off to hospital every time I felt a little poorly, but he quickly realised once he'd said this that I was so weak that I was actually struggling to stand up and he rushed to support me, retracting his earlier comments with his actions.....

I actually felt so light headed and spaced out that I described myself to my husband as feeling like a puppet on a string and that when my brain was telling my legs to move in order to walk there was a delay between the messages leaving my brain and my legs actually moving. I had to use huge physical effort just to lift my feet and when I did I felt like I was walking on air or *"moon walking"* as I could barely feel the ground underneath me. It was an almost out of body experience. I started to sob, but I was so dry that there were hardly any tears coming out of my eyes, I was just heaving with despair.

As I was too weak to drive, my husband drove me to A+E at the NHS teaching hospital where both my surgeon and gastroenterologist have their NHS posts and due to my extensive history and visible weakness I was triaged within about fifteen minutes. At triage, when my blood pressure was taken it was quite low (90/40) and so I was put on IV fluids straight away for dehydration, given some pain relief and antiemetics and also sent for an abdominal x-ray. The x-ray showed some dilated small bowel loops on the right which were once again suggestive of another sub-acute small bowel obstruction. Needless to say I was totally gutted because it felt like *"groundhog day"* all over again.

I was admitted to a ward overnight and was seen by a consultant who assured me that as I was already a private patient of his colleagues I would be reviewed by one of them on the Monday. However, by midday on Monday neither my surgeon nor my gastroenterologist had been to see me and as I was starting to feel a bit better in that the pain had eased and I felt more hydrated I was wanting to get home so I could get to work the following day, which was my planned return to work day following the festive break.

(I'd already been absent from work for almost twenty six weeks in the previous year and when I had been able to work I was working reduced hours and days. I was therefore starting to worry about losing my job or at least having to reduce or vary my work pattern as my health simply wasn't good enough to work the hours I was contracted for.) I therefore telephoned my surgeon's secretary and booked a private appointment with him for the following night with a view to discharging myself, that evening, attending work the following day and then discussing things with my surgeon the following evening.

By six thirty on the Monday evening I'd had three litres of IV fluids, was feeling a bit stronger and my husband had just come to visit. And as I'd still not seen my surgeon or my gastroenterologist I told my husband I wanted to go home with him because I was so anxious to get into work the next day. So I called my assigned nurse and discussed my discharge with her. I explained to her that I was seeing my surgeon privately the following night and that I just wanted to get home. She said that because I was being followed up privately I could go, but on the understanding that I came back if I felt any worse and that I would have to sign a disclaimer and discharge myself because it was not in my care plan for me to be discharged yet. I therefore signed the disclaimer and went home with my husband that evening. And whilst I was glad to be back home and to be seeing my girls I was still in pain and still felt nauseous.

I had quite a fitful sleep that night and when I woke up the following morning, again drenched in sweat, despite not feeling well enough, I decided to attend work. After all, I'd just had two weeks holiday over Christmas and I was determined to throw myself into things. I'd been assigned to a high profile three month project just before the Christmas break and having work to focus on and return to was a real lifeline for me in terms of getting my life back to normal.

When I got into the office the following day I didn't bother telling anyone in work that I'd been in hospital on IV fluids the day before and just got on with my day – which I really enjoyed apart from the fact that I felt completely wiped out by mid-afternoon. Then by the time I had driven the hour and fifteen minute commute from work to my appointment with my consultant I was feeling so dehydrated that my mouth felt like sandpaper, and so nauseous that I had absolutely no appetite.

At my appointment my consultant ordered a further x-ray to see if the dilated bowel loops had settled and as things hadn't completely subsided he suggested that I be admitted into hospital again overnight for some more hydration. He said that if I felt stronger in the morning after a further two litres of IV fluids I could go home and attend work the following day (Thursday) but that I should come back for a further review on the Friday. I also raised the issue of being able to contact him when I got poorly because I felt that after everything I'd been through under his care he should at least make himself more available to me. I didn't directly ask for his mobile number or email address because I was conscious not to overstep the doctor/patient boundaries, but I didn't have to. He gave me full contact details should I need to get in touch in an emergency.

Once I'd been admitted and taken to the ward and my fluids up I asked the night staff for a sleeping tablet as I'd not slept well for the previous few nights. And with the sleeper I actually managed a really good sleep, despite the IV line in my arm.

When I awoke the following morning I felt a bit better but because my blood pressure was still very low my nurse called my consultant to query my discharge. My consultant prescribed another half-litre of fluids and said that I should not be discharged until my blood pressure had stabilised. By mid-afternoon on the Wednesday after the full two and a half litres of IV fluids my blood pressure had come up and I was discharged. I went home, not feeling brilliant, but stronger than I had done on my admission the previous evening.

I then attended work again on Thursday but by mid-morning I couldn't concentrate, I was nauseous, dizzy, and had an unquenchable thirst. I realised that I was still dehydrated and I had started to feel like that puppet on a string. I knew I needed to be re-admitted to hospital and so I went to speak to my line manager. I told him I'd been in hospital twice that week already for IV fluids and he told me to leave work and get home, but only if I felt well enough to drive the longish journey. I assured him I was well enough to drive, whilst in reality knowing I probably wasn't – but I couldn't face the thought of being admitted to hospital over an hour away from home.

The journey absolutely finished me and I had a near crash at a roundabout shortly after I'd left my work site. And by the time I arrived home I was so weak that I literally collapsed onto my sofa. After a few minutes I managed to summon up enough energy to call an ambulance, which then arrived fairly quickly and took me to my nearest hospital A+E.

On arrival at the hospital the paramedics in the ambulance handed me over to the consultant who then took my history. I explained to the consultant that despite me having been hospitalised twice in the preceding few days and receiving IV fluids I still felt nauseous and dehydrated. My blood pressure was still low and my heart was racing and my muscles had started cramping up. I was also in so much pain that I couldn't stay still. But because of my dehydration I had an unquenchable thirst and dry mouth and so had a bottle of juice with me from which I was taking small sips to keep my mouth moist. And when the consultant handed me over to a more junior doctor to take my bloods and cannulate me I heard the junior doctor say:

*"but she's drinking"*

in response to the Consultant when she said I was dehydrated and needed fluids.

Once my bloods had been taken and the cannula inserted, the junior doctor asked me to try and pass a urine sample if I could so that it could be tested. I managed to pass a small amount of urine in a sample pot and gave it to her and because the sample wasn't deep yellow the doctor said that from my urine I didn't appear to be dehydrated. The junior doctor then went on to say that she was giving me fluids and antiemetics because I had asked for them and the consultant had prescribed them but that in her opinion she felt that I didn't need them and she questioned whether or not I was just having a panic attack because I'd been through so much.

When she said this I did immediately start to question my own sanity and began to wonder whether the pain, nausea, dizziness and feelings of weakness were all in my head....... I was also aware that I had an appointment scheduled with my consultant the following night and as I didn't really want to be in hospital in any event I told myself that after the litre bag of fluids had run though I would just ask to be discharged. I reasoned that maybe the junior doctor was right and perhaps the reason why my heart was racing was because I was panicking and not because my heart was beating faster and harder to try and keep my blood pressure up. So, even though I still felt quite weak after the litre of fluids had run through I asked the nurse if I could go home and I left A+E and called a taxi to collect me.

I didn't feel too bad that evening as the fluids had obviously helped a little and as I was less nauseous I managed dinner with my husband, followed by a decent night's sleep.

But by the time I woke up the following morning I was feeling very weak again.

Nonetheless I forced myself to get up and get on with things. So I dropped the girls off at school and then popped to the shops to buy a few essentials. But as I was walking around the aisles of my local supermarket I felt incredibly dizzy and lightheaded. I felt like that puppet on a string whose arms and legs weren't responding very quickly to the strings on the sticks.

I finished my shop early without getting everything on my list (because I was simply physically unable to continue shopping) and just about managed to drive myself home. I then called my mum, who on hearing how I felt was really concerned. She said that she'd like to attend the consultant's appointment with me and my husband which was scheduled for that evening and so she drove over to be with me. When my Mum arrived she commented on how frail I looked and also on the big dark circles under my eyes.

I was feeling too weak to do anything by this point and my mum therefore collected the girls from school for me and made sure I just rested until my appointment later that evening.

By the time I arrived back at the hospital for my appointment with my consultant that evening I was in a considerable amount of pain and very dizzy, lightheaded, thirsty and nauseous. My consultant advised me, my mother and husband that I needed to be re-admitted for some further hydration and pain relief that evening and for an MRI scan the following morning to further check my small intestine to try and elucidate the cause of the pain. He also advised that he would now be referring me to a physician, who specialises in fluid imbalances to see if he could try and sort out my continuing problems with dehydration.

I remained in hospital over the weekend and was given a further eight litres of IV fluids and was then discharged on the basis that the MRI scan was clear and that I would be seeing this general physician during the following week.

I met my new consultant for the first time on the following Tuesday as an outpatient at the hospital where I'd had my colectomy almost a year ago to the day. He explained to me during the consultation that I was suffering from hypovolemia, namely a decrease in the volume of blood in my body due to salt and water depletion and that he would need to re-admit me to hospital for a series of tests and to bring me back into water balance. He said that he would do this over a period of days by getting my medication right as opposed to giving me IV fluids which, in his words, was just:

*"papering over the cracks".*

My physician explained that due to my excessive diarrhoea, the electrolyte and water loss from my ileum was cumulative and over time had led to depletion in my circulating blood volume. This in turn had led to the episodes of near collapse from which I'd been suffering. He further explained that the five sachets of double strength dioralyte treatment that I was on were insufficient to enable me to maintain normal body physiology in the context of my intestinal losses.

The ease with which this physician diagnosed my condition was a real relief but it also made me feel let down by the gastroenterologist, whose care I'd been under the previous summer and who didn't manage to relate my symptoms of nausea and light-headedness to dehydration alone.

Instead he carried out the tests for lactose and glucose intolerance and did a gastroscopy and because these tests revealed nothing he simply said that there was nothing else wrong and didn't even think to revisit my hydration status by increasing my dioralyte or limiting my other fluids.

I knew that I wasn't right at that time but the gastroenterologist didn't listen, or if he did, he didn't have the skill set to get me well. I really disliked this gastroenterologist because his bedside manner was to start off from the stance that there is nothing wrong until he proved that there was. So when my tests came back negative he would dismiss me like a hypochondriac.

I also felt a little let down by my surgeon, albeit to a lesser extent. He should really have pushed his gastroenterologist colleague harder for a diagnosis the previous summer or referred me elsewhere at that point. I'd clearly been suffering with hypovolemia since my colectomy the previous January but as I was being re-admitted so frequently, on average every three to eight weeks for the obstructions, stricture dilatation and then the internal hernia, and being given IV fluids on each admission I was managing to sustain myself,……just.

So, my new physician re-admitted me to hospital on the Friday following my initial consultation with him and when he came to see me that afternoon he said that he wanted to get me back to euvolemia by means of oral medication alone.

Prior to becoming a medical doctor, he explained that he was a doctor of chemistry and that he was determined to get a solution for me without having to resort to IV fluids. I wasn't particularly encouraged during this conversation because I felt like I was being treated as a chemistry experiment or a *"thing"* with a chemical imbalance and not a real human being with a complex health issue. Nonetheless I put my trust in him because I was clearly so unwell.

I'd been weighed on my arrival at hospital that afternoon and my weight was fifty seven point eight kilos but by the next morning it had dropped to fifty seven point three kilos and so my physician's first bit of advice was to tell me to drink more fluids (at least two litres) to replace what I was losing and that he would be increasing my dioralyte to ten sachets a day to replace the electrolyte losses.

But rather than the double strength dioralyte that the dietician had prescribed he would be giving me single strength which would mean I had to consume a further two litres of fluids. I was therefore told to consume at least four litres of fluids a day.

My physician reasoned that the double strength dioralyte was very salty to taste and that drinking that was enough to be making me feel thirsty and that by making it single strength it would be more palatable. (He either ignored, or was blind to the fact, that the sodium concentration of double strength dioralyte at 120mmol/L aided its absorption.)

Anyway, I did as I was instructed and increased my fluid intake and the number of sachets of dioralyte – albeit in a reduced strength and by the Saturday evening I was feeling so poorly that I was fighting to stay conscious, even lying down in my hospital bed.

My mouth was so dry that I felt like I was spitting feathers and I felt incredibly weak and shaky and was in excruciating pain in my pelvis and lower back. I called my nurse and asked her to contact the RMO and told her that if I were at home I would take myself to my nearest A+E because I felt that I was very dehydrated – to the point of near collapse and needed some IV fluids.

Before she contacted the RMO she took my blood pressure and heart rate and could see that my blood pressure had dropped dangerously low and my heart rate was very high. When the RMO arrived she could also see that I was very distressed and she asked if I was anxious and whether or not I wanted any diazepam to calm me down?!?!?

I told her quite flatly that I didn't want anything to calm me down, that I wasn't anxious but that I really felt like I was so dehydrated that I was fighting to stay conscious. She explained to me that she could see that I looked dehydrated but that my consultant had given strict instructions that I was not to be put on IV fluids. I explained that I didn't care, that I knew that I needed them and that she had to contact my consultant for him to come and review me himself.

The RMO contacted my consultant and then came back to see me and said that his instructions remained the same and that I didn't need any fluids and that the whole purpose of me being in hospital was to render me euvolemic without using IV fluids.

By this time I was crying with a whimper out of sheer frustration and pain but my eyes were so dry that no tears were even coming out. I was really at my wits end and called my husband and asked him to come and collect me to take me to the nearest NHS A+E.

My husband was as amazing as ever on the phone and helped calm me down. He suggested that I send a text to my surgeon who had referred me to the physician to see if that made any difference. So I sent a text saying that I was in desperate need of fluids, that I felt like I was about to pass out and that if I didn't get IV fluids I was going to discharge myself and get my husband to take me to the nearest NHS A+E.

The text did the trick. My surgeon didn't respond to the text but within the hour I was back up on IV fluids and so he must have reached out to my physician and asked him to put me back on the fluids. By the following morning my weight was back up to fifty seven point five kilos and I had started to feel a little better.

When my physician came to see me the next day he said that he wasn't happy about having to put me back on the fluids and that he would wean me off them. The way he spoke to me was almost as if he thought I wanted to be on the fluids and wanted to be in hospital …..it was very demeaning.

Anyway I just had to grit my teeth and put my faith in his ability….

This in-patient stay lasted for twenty two days, during which time my medication was changed significantly.

I was taken off dioralyte (which my physician felt was contributing to my thirst symptoms) and instead prescribed "*slow sodium*" tables to aid fluid retention, loperamide to reduce my diarrhoea and Ascorbic Acid to neutralise my body's PH balance. I was also prescribed colestyramine to further reduce my diarrhoea, because despite being put on the highest doses of loperamide I was still losing too much fluid.

For some reason, and as I'd already previously found and told my physician, the loperamide didn't work well in terms of bulking my stool up.

My physician explained to me that he thought I had a "*bile acid sensitivity*" which was preventing my small intestine adapting following the colectomy. Indeed he explained to me that most people after a colectomy will adapt in that whilst their stools will not return to normal they may only have two to four bowel movements a day and will not become hypovolemic.

Unfortunately, however, I have not adapted following the loss of my colon and will thus have this lifelong propensity to hypovolemia. If left untreated, hypovolemia will lead to hypovolemic shock and eventual death……. Wonderful.

So, back to the mad chemist's treatment plan…..

Whilst the colestyramine initially seemed to work by bulking up and reducing the number of my stools, after several days I returned to having around twelve bouts of diarrhoea a day and all that was preventing me from dehydrating was the IV fluids (when I was on them) and the very high dose of sodium chloride I was prescribed.

Indeed by the time of my discharge I was taking four hundred and eighty millimoles of sodium chloride a day (as forty eight slow sodium tablets) which is just over the equivalent sodium chloride found in three litres of Baxters IV solution (four hundred and sixty two millimoles).

I was also advised to drink at least seventy millilitres of fluid for each tablet taken, so that was at least three thousand three hundred and thirty six millilitres or three point three six litres of fluid a day, but all that seemed to be happening was that the more I drank the more fluid I would lose either as urine or diarrhoea.

I was also a bit worried about being prescribed so much sodium chloride because the medication label itself said that the absolute maximum that could be taken at any time was twenty tablets and also because it was *"slow"*.

I was concerned that I wouldn't actually be absorbing much of the sodium chloride anyway because my transit time was so quick. I'd already known that I wasn't absorbing the venlaflaxine I used to take for my anxiety symptoms and had subsequently been switched to duloxetine, but when I mentioned the dose to my consultant and my quick transit time he just told me not to worry and that I could still also add salt to my food and for me there was never going to be any such thing as too much salt.

During the stay, whilst the loperamide and cholestyramine didn't slow my diarrhoea down to any useful degree, what it did do is thicken my stool enough for me to suffer from excruciating pain at the site of my anastomosis when passing stool. I also developed chronic pain up my rectum and in my lower back, particularly if my rectum was filled with stool. I therefore had another defecating proctogram to check whether there was a recurrence of the prolapse because I felt the pain in the area of my sacrum (which is where my rectum had been fixed at the time the prolapse was repaired).

The proctogram showed that there was no visible anatomical abnormalities any more (i.e. the surgery to correct my prolapse had been successful) but I couldn't actually feel anything when the barium paste was inserted into my rectum and the radiologist commented that I obviously had a loss of sensation which was more than likely due to pudendal nerve damage.

I knew when he mentioned this that the nerve damage must have been caused during my second labour because following the birth of my second daughter I'd never really felt the urge to defecate in the same way as I had before. Further, I knew from the labour itself that my little girl had been stuck at the level of the ischial spines, behind which the pudendal nerve is located.

Having the radiologist tell me this was no great surprise because I already knew that I'd sustained injuries during that labour..... but it was still depressing to hear ....

This in patient stay was also depressing from the point of view that during the stay my physician was careful to manage my expectations of my health and its impact on my life moving forwards. In particular, he made it very clear that keeping my body in both fluid and PH balance was vitally important and that a large fluid imbalance would lead to hypovolemic shock and could be life threatening. He felt that due to the careful balance I have to achieve and my propensity for fatigue, dizziness, lack of mental sharpness, thirst and nausea when my fluid levels drop meant that travelling long distances to work on a daily basis would just not be feasible.

He suggested that I needed to find work closer to home. He also said that I shouldn't expect my health to return to where it was before I became ill. He explained that the function of the colon is to maintain water balance in the body and without it, because my bowel movements were so watery and so frequent, I would always struggle to maintain balance and would have this propensity to hypovolemia. My physician said that he would not be able to achieve "*perfection*" through medication but that he hoped he would be able to get me to a stage where I could become as hydrated as possible and then maintain that status.

The other thing that my physician pointed out was that I could never expect to feel completely pain free because I would inevitably have adhesions from the sheer amount of surgery I'd had and that these adhesions would cause pain. He made it clear that I just had to accept these limitations and make alterations to my life accordingly.

As you can imagine, hearing this about my long term prognosis was pretty devastating and being told I'd need to get a job closer to home just added to the blow. His clear implication was that my fluid and PH imbalances were a disability and that my career aspirations needed to change accordingly. I'd already had to forgo a job promotion the previous year when I realised that I was getting really poorly (and therefore missed out on the increased responsibility, higher profile and significantly better package that would have gone with this), so to re-adjust again to lower my expectations further in terms of my future career wasn't easy.

I was thirty nine at the time and had a proven track record as a successful commercial and intellectual property lawyer with great career experience (in both private practice and industry), so being told by my physician that my health limitations were such that I would need to take a back seat with my career was a really tough pill to swallow.

But what irked me most about this stay in hospital was that I was really worried that the regime my physician had put me on was completely wrong for my condition and I had serious reservations about whether I would ever feel properly well again.

In terms of the regime, swallowing forty eight slow sodium tablets a day just seemed plain wrong. Then one morning my assigned nurse came to give me my morning medication and promptly queried the slow sodium dose. She said:

*"are you sure you're due twelve slow sodium tablets four times a day? Let me just go and check that."*

I assured her I was on forty eight tablets a day and she just raised her eyebrows as if to say *"that's really odd"*.

She then proceeded to count the tablets out, still with markedly raised eyebrows whilst explaining to me that she used to work on an intestinal failure unit and that she had never heard of anyone with intestinal fluid losses being prescribed high doses of slow sodium. She also asked how much I'd been told to drink and when I said that I had to drink at least seventy millilitres of fluid for each tablet (so three thousand three hundred and thirty six millilitres or three point three six litres of fluid a day) she expressed real concern that my physician had asked me to drink so much and take so many tablets.

She said that people with a short bowel needed to drink less fluid and not more and drink an electrolyte mix.

I wasn't particularly happy when she told me this because as a patient I'd put my faith in my physician, but I instinctively knew that what she was saying made sense. I had previously read about marathon runners *"washing"* salt out of their bodies and teenagers who'd taken ecstasy dying because they also drank too much water.

I therefore did some research on intestinal failure and the recommended fluid balance regimes and realised that the leading teaching hospitals in this area recommended fluid restrictions and electrolyte drinks; **NOT** drinking freely (and at least three point three six litres a day) and consuming forty eight slow sodium tablets as I had been advised…

So after twenty two days under my physician's care and in serious doubt about the medication he had placed me on I asked to be discharged and went straight to my (new) GP the following day. I told her about my in-patient stay and my worries about my physician's advice and asked to be referred to a consultant who specialised in intestinal failure.

My GP was very understanding and thought that it was certainly a sensible step to be taking given that the doses of slow sodium I'd been prescribed well exceeded the recommended maximum dose and were therefore unlicensed doses.

I was initially referred to a consultant gastroenterologist who practised at one of two Intestinal Failure Units in the country and which was relatively close to home, but as he'd stopped seeing private patients it meant that I'd have to be seen through the NHS which would have taken weeks.

It was already mid-February at this point, so over thirteen months after my colectomy and I was still feeling too weak to return to work. I therefore went to see one of his NHS colleagues privately (who whilst also a gastroenterologist, didn't specialise in intestinal failure) in the hope that I could get an internal NHS referral and thus an appointment more quickly. This gastroenterologist was lovely in that he listened carefully and confirmed that he felt the regime I'd been put on wasn't right; he also advised me to go back to ten sachets of double strength dioralyte a day and referred me to his colleague requesting a period of in-patient assessment.

Unfortunately, however, this did nothing to speed things up and after a month had elapsed following my initial appointment I reached out over LinkedIn® to the gastroenterologist I'd been referred to in order to request an urgent appointment. I also asked my GP to send him a further urgent referral. I was getting incredibly frustrated with the status quo, particularly because my employer was becoming increasingly restless with my prolonged absence and home working and I knew there was absolutely no point whatsoever trying to return to the office before I was established on an adequate regime……..

Anyway, whether it was the LinkedIn® message or the further contact from my GP that expedited things I don't know, but I was given an outpatient appointment within days.

The appointment was, however, rather disappointing. I was seen initially by a surgeon rather than the physician to whom I'd been referred. And whilst the surgeon took a detailed history, he really didn't give me any solid advice other than to re-iterate that I should stay on the dioralyte rather than the slow sodium and:

*"try not to drink too much other fluid"*

He also said that he felt that my problem was not strictly intestinal failure *per se* because I still had the full length of my small bowel and that because there was such a long waiting list for an in-patient assessment in the intestinal failure unit things could be better expedited by the gastroenterology team who had a shorter waiting list. He did, however, say that the physician to whom I'd been referred wanted to speak to me before I left. So I met the physician but was singularly disappointed.

He'd obviously spoken with the gastroenterologist I'd seen at my local private hospital the preceding summer and with whom I'd also been disappointed by and he also knew the physician under whose care I'd been that January and who put me on the slow sodium regime. And rather than spend any time with me he simply said that his surgical colleague had spent time with me and given that I'd recently had a prolonged in-patient assessment he queried why I felt I needed another one. I simply told him that I didn't feel well and that I still felt like I was suffering with dehydration and that I needed someone with appropriate expertise to put me on a regime for my optimum health moving forwards and to rule out any other reason for my continued diarrhoea.

I explained that one of my concerns was that I wanted to know why I was getting so dehydrated. It just didn't make sense to me that my dehydration was simply due to my intestinal losses and I instinctively felt that I mustn't be able to absorb my electrolytes properly. I recounted that during the caesarean section for the birth of my second daughter my intestines were temporarily removed from my peritoneal cavity and were placed outside of my abdomen and looked grey in colour, thus indicating ischaemia. (Explaining that my dad had commented on this as he was looking through the photographs my husband had taken at the birth ....)

I further explained that as I wasn't getting any answers (aside from the fact that my losses were too high) I had done some internet research myself and had found an article entitled:

*"Response of the intestinal mucosa to ischaemia"* (*Gut*, 1981, **22**, 512-527)

which explains that the response of intestinal mucosa to ischaemia is that during the course of regeneration water and ion absorption is retarded and the villi are considerably shorter than normal. The net effect of this is that absorption of water and sodium is adversely affected. I therefore postulated that one explanation for my condition could be that whilst I still had my colon, this failure to absorb was not a problem because the colon did the absorbing; but as soon as my colon was removed the failure to ensure net absorption of water and sodium became problematic.

The physician said that he thought this was unlikely because the bowel repairs itself quite quickly but that he thought I should see one of his gastroenterologist colleagues outside of the intestinal failure unit to be assessed in any event. I thanked him for his time, but given that over six weeks had elapsed since my discharge and I was not feeling any better I felt like the trip had been a waste of time. I was basically given no real advice and simply told that I would need to come back as an outpatient to see this gastroenterologist colleague who may then admit me for a period of in-patient assessment.

I therefore went straight back to see my GP who then referred me to the lead gastroenterologist at the other centre for intestinal failure, which is over one hundred and seventy five miles from my home. Fortunately this consultant also ran a private practice at the same hospital where his NHS post is located. Initially I hadn't really wanted to travel to see a specialist because it made no sense whatsoever for me to have to travel over three hours to be hospitalised when I became ill with dehydration, but as I was not really getting anywhere quickly on the NHS closer to home I reasoned that I had no option but to bite the bullet and go privately further afield.

And it was certainly worth it. I was seen privately within three weeks of my referral and the consultation was a breath of fresh air. My husband drove me to my appointment and accompanied me through the consultation and he too was incredibly impressed with the consultant and the facilities.

In particular, the consulting rooms were located next to a private ward on the top floor of the NHS hospital where this consultant had his NHS practice which meant that he was always on hand for both his private and NHS patients. This set up made far more sense than the situation back home where my private hospital is several miles from the main NHS teaching hospitals in the area which inevitably means that the consultants only tend to visit late in the evenings after their NHS work ends or in between their clinic lists at the private hospital…..furthermore it made absolute economic sense for the NHS because the private wing makes use of the NHS facilities including operating theatres, pathology services and imaging services therefore creating a valuable revenue stream …..

But what set this consultation apart even more was the attentiveness with which the lead gastroenterologist listened. He was clearly at the top of his profession to hold the post he had but he was humble and personable and wasn't afraid to say when things are not entirely understood.

Indeed when I asked him whether he thought that my small bowel could have been permanently damaged due to the ischemia during my second caesarean section, whilst he expressed that he thought this was very unlikely because the bowel tends to repair itself quite quickly, he didn't dismiss the idea out of hand. He simply said that a more likely explanation was that I was simply losing too much fluid because of the loss of my colon (whose job it is to absorb water) but that this would have to be assessed.

He commented that with the amount of diarrhoea I was having it was likely that I would be unable to replace the electrolytes I was losing with the slow sodium I had been prescribed, because it is usually not effective in patients with diarrhoea precisely because it is "*slow*". He also explained that drinking the amount of water I had been told to drink would actually make me more dehydrated and not less dehydrated.

I was therefore immediately established on a short bowel regime which entailed a combination of fluid restriction and oral rehydration therapy. Rather than continue with the dioralyte, which is really designed for rehydration of short term secretory diarrhoea caused by infection such as cholera, I was prescribed a litre of an electrolyte mix per day which contained a litre of water, one teaspoon of sodium chloride (table salt), six teaspoons of glucose and half a teaspoon of bicarbonate of soda (**E-Mix**) and told to restrict all other fluids to one litre a day.

My gastroenterologist explained to me that as I didn't have a colon, most of the fluid I'd been drinking previous to being put on the short bowel regime would have been passed straight out of my body flushing sodium with it. He explained that ideally he needed to slow down my bowel habit and also establish me on the fluid restriction and the E-Mix which was at the optimum concentration for absorption by my small intestine.

I explained to my new gastroenterologist that the problem with slowing my bowel habit was that when my stool thickened up slightly with the loperamide I suffered from increased rectal pain at or near to the site of my anastomosis, together with increased abdominal pain. As such, I explained that I'd made a conscious decision that I would prefer to suffer from diarrhoea and have a propensity to dehydration, rather than be in pain every day. I really didn't want to have to take heavy duty painkillers for the rest of my life…..

Perhaps the most important and fundamental thing he explained to me though, was that in dehydrated patients serum levels of potassium and sodium (i.e. the levels in your blood) can appear normal on blood tests because the dehydration itself has led to a decrease in circulating blood volume rather than a lowering of the electrolytes.

As such, the most sensitive test for dehydration is a urinary sodium test. He said that this could either be done by a twenty four hour collection or just a random sample. He therefore arranged for me to have a random urinary sodium test just after the initial consultation and explained that if my urinary sodium fell to less than 20mmol/L that this is an indication of dehydration and it is at that point I would need parenteral (IV) fluids.

When he told me this together with stressing how insensitive blood tests were for dehydration I was absolutely gobsmacked that none of the medics (and especially the gastroenterologists) who I'd already been seen by had informed me of this or carried out a random urinary sodium test.

I did have one sample tested shortly after I'd seen the dietician the previous summer but because on the one occasion I had it tested my levels were over 20mmol/L I didn't think to get it tested again, but more alarmingly, the gastroenterologist who I'd seen privately as an in-patient in my local hospital didn't once order this test when I complained of being so dehydrated that I felt too nauseous to eat and too weak to be discharged. Instead he simply tried to convince me that I had just been through a lot and would probably feel better at home….so having my new gastroenterologist explain this simple test was like a Eureka moment. He also arranged for me to have a litre of saline IV before I went home to help tide me over to my in-patient assessment.

And sure enough when I was admitted later that month for my in-patient assessment my urinary sodium was only 12mmol/L and it dropped again during the admission. On each occasion when my sodium levels dropped I was given IV fluids to get me better hydrated.

In addition to my gastroenterologist I saw the senior biofeedback nurse and lead colorectal surgeon. I also underwent a whole battery of tests during my stay so that my new gastroenterologist and his team could ascertain if there was anything further that he could do to help me. And I certainly felt that I was in very safe hands at all times and was treated with respect, not once being made to feel like I was a *"head case"*.

One of the tests I had as an inpatient was anorectal physiology which entailed a probe being inserted into my rectum to test the sensitivity and the integrity of my sphincter muscles. In short, the upshot of the assessment was indicative of pudendal nerve damage, confirming the earlier diagnosis of nerve damage following my second defecating proctogram. The daily urinary sodium tests and fluid output charts established that I had a high output and that my gastrointestinal losses were such that I would inevitably end up in negative fluid balance from time to time.

A barium follow through that I had during this stay showed that I had prominent bowel loops in my pelvis when pressure was applied thereby indicating adhesions and a flexible sigmoidoscopy revealed that there was no stricture at the anastomosis, but what I did identify from this test was that I felt pain at the site of my anastomosis and that when I get rectal pain it is at that point.

I knew this because even though I was sedated for the procedure I could still feel pain and when I asked my consultant where he was in my bowel he confirmed that he was at my anastomosis.

A proctogram also revealed that there was no recurrent rectal prolapse and an MRI of my lumbar spine showed that there was nothing in my spine that would be causing the low back ache and deep pelvic pain I suffer with from time to time and in particular when I start to become dehydrated.

When I met the surgeon she explained that because there is a risk when operating on adhesions that they will just reform following the further surgery, she instead referred me to the lead biofeedback nurse who suggested that as SNS had previously helped me with pain, a percutaneous tibial nerve stimulation device may give me some relief from my symptoms.

She said that the tibial, pudendal and sacral nerves are all nerves of the sacral plexus and that by using tibial nerve stimulation I might get some relief from the chronic pelvic pain.

The device itself is a bit like a tens machine with four electrode pads, two of which are placed on the inside of each ankle with the other two being placed a short distance above. The wires then connect to the device itself which has controls to increase and decrease the frequency to the electrode pads. I used the device for thirty minutes each day during my stay and did notice improvement in my pain and so I was sent home with my own device which I used almost daily following my discharge.

Without doubt, the most useful outcome of my inpatient stay (aside from the tibial nerve stimulation device) was that I was told that the truest test of whether I am dehydrated is to check my urine sodium and that if this falls to below 20mmol/L (normal range is 20-240mmol/L) I will likely need two to three litres of IV fluids to bring me back to fluid balance. This small piece of useful information has now enabled me to enjoy a far better quality of life because I have not become very poorly again with hypovolemia.

Instead I deal with the dehydration as soon as it starts to set in. So, since my discharge, whenever I feel dry I have my urine sodium checked by my GP and if this is below 20mmol/litre I contact my consultant near home who then admits me to my local private hospital overnight for rehydration.

It was so worthwhile travelling to see my new gastroenterologist as I was given a thorough going over and put on the right regime and a plan for when I become dehydrated. He also explained that when a patient becomes dehydrated they can become symptomatic with a sub-acute small bowel obstruction because of the fluid losses.

This information has helped me become calmer when I become dehydrated and have obstructive symptoms because I now know that in all likelihood the obstruction will resolve with the IV fluid therapy. My gastroenterologist also ensured that my local consultant surgeon was copied in on my care plan so that I would be given continuity of care whilst at home and I have also been given a letter (which I have uploaded to my documents cloud) which I can show to any doctor (anywhere in the world) and which tells them what to do if I become dehydrated.

These relatively small but immensely important measures have given me such piece of mind and enabled me to live a more carefree existence.......

# My altered normality

So normality for me now is drinking up to a litre of water a day mixed with six teaspoons of glucose, a teaspoon of salt, half a teaspoon of bicarbonate of soda and a bit of lime juice (to make it more palatable) and a fluid restriction of around one litre of other fluids.

Perhaps surprisingly I find drinking the E-Mix the easy bit and the restriction the hard bit. A litre is only four cups of tea or a couple of cups of tea, a glass of juice and a glass of wine and so whilst I try really hard to stick to the restriction I don't always manage it, particularly if it is hot and I feel thirsty or when I'm at social functions.

Then when I feel like I'm getting dry I get my urine electrolytes checked and if I am dehydrated I get admitted to hospital for IV fluids.....

From getting my urine sodium results to actually getting the IV fluids up and running usually takes around four to five hours, largely because a bed needs to be found for my admission. So I'm usually not admitted until early evening the day after I've dropped my sample off at the GP or the local pathology lab. I'm then given around three and half litres of normal IV saline (sometimes with potassium added if my potassium is also low) during that night and the following day.

One of the worst things about getting dehydrated is that for some reason my pain increases and I feel incredibly nauseous, to the point where I can't tolerate food (especially when my potassium is also low). I can also become lightheaded to the point where I feel unsteady on my feet and am in danger of passing out, but thankfully over time I've managed to know the early warning signs and during the first summer of my new *"altered normality"* I required admissions on 8 July, 16 August, 24 September and 29 October.

I was hospitalised overnight each time and promptly discharged the following day feeling revitalised .... except for my hospitalisation on 29 October.

When I got my urine results back on 29 October my sodium level had dropped to its lowest possible detectable level of <10mmol/l. So I called my consultant's secretary to arrange for admission, but unfortunately my usual consultant was on annual leave, and so I was admitted under the care of one of my consultant's colleagues.

On this occasion, whilst I was feeling weak with dehydration, I also had excruciating left flank pain and was worried that my kidney was starting to play up with all my episodes of dehydration. It was because of this pain that I was first sent to the ECC to get assessed by the ECC doctor, rather than being sent straight to the ward for fluids. During the examination the ECC doctor told me that he wanted me to have a scan because of the pain in the area of my left kidney. So I was taken to the imaging department where the scan revealed a small cyst in my left kidney and also some *"hyperperistalsing loops of small bowel in the upper left quadrant which could be indicative of a developing small bowel obstruction"*.

When I returned to the ECC, the ECC doctor explained that the kidney cyst was unlikely to be significant and that the hyperperistalsing loops of small bowel were the more likely culprits for the pain. I was therefore ordered to be nil by mouth for the following twenty four hours whilst being administered the IV fluids in the hope that during this time everything would settle down.

By mid-afternoon the following day I still didn't feel great in terms of my hydration status (despite the IV fluids) and I was still in pain, but as it was half term and my elder daughter's ninth birthday the following day I decided to go home anyway despite the pain. (I'd had to take oramorph for the first time in well over a year during this admission and I had to be discharged home with some because my pain was so bad.)

Even though I was poorly, we still managed to have a great day for my daughter's birthday. My eldest is a *"Halloween baby"* and so she always gets dressed up to celebrate and this particular year, rather than have a party, she'd decided she wanted to take her best friend, together with the rest of the family, to a *"haunted old hall"*, followed by dinner in her favourite restaurant, some trick or treating around the village and then a sleepover.

The girls had a fantastic time but I really struggled throughout the day and into the evening with nausea, dehydration symptoms, deep pelvic pain and left flank pain. I dosed myself up with painkillers and was as careful as ever not to complain about my pain in front of my girls, but my husband could see that I was noticeably weak and lacking in sparkle and energy….I even struggled to eat my favourite main course and the birthday cake at dinner – which is not like me. Anyone who knows me well will vouch for the fact that I can quite easily eat cake at any time of day or night, including for breakfast!

The following day I'd arranged to take the birthday girl to see *"Riverdance"* at the theatre as a surprise followed by dinner with the family and her Grandma. My eldest is a keen (and very good) ballet, tap, jazz, street and modern dancer and I knew she would love the sound of the shoes on the stage. And she did. Especially the *"dance off"* between two male Irish dancers and two male tap dancers. She was so thrilled with the outing (even though she would actually have preferred to see *"One Direction"* had they been performing that day) and kept squeezing my hand and kissing my cheek to thank me.

Then about half way through the performance I started to get agonising abdominal pains, deep in my pelvis and through to my rectum and even my little girl could see the pain etched on my face, which I was simply unable to hide. And the pain didn't abate. We still went to Grandma's for dinner though once the performance had finished because, as usual, I didn't want anyone to miss out. So I put on the bravest of faces whilst everyone else was happy and pain free, including my incredibly fit and glamorous seventy one year old mother.

By the time dinner was served I was feeling really nauseous again and didn't manage much food and as soon as everyone was finished I excused myself from the table and went to the bathroom. When I wiped myself I realised there was fresh bright red blood on the paper and then when I looked in the bowl the entire stool and toilet water were coloured red.

It was just after eight in the evening on the Saturday night when I first passed blood and I had two more blood coloured stools before I went to bed. Then, when I woke for a bowel movement about three the following morning I was still passing fresh blood with my stool, so I woke my husband and told him that I was going to take myself off to A+E.

I was conscious that it wasn't a small amount of blood that I was losing and with my history I knew that I couldn't be too careful. But I was also dreading having to visit a busy city centre A+E in the small hours of Sunday morning.....

I called a cab to take me as I wasn't up to driving, and thankfully the cab came quickly. On arrival, the waiting area was largely full of people who'd either had too much to drink or taken too many drugs and were nursing predominantly self-inflicted injuries. I was, however, triaged quickly and then taken to majors where my blood was taken.

But it then took what seemed to be an age to get me any strong pain relief and anti-emetics and to get my fluids up and running. I had several more bloody stools whilst in A+E and was frustrated in the extreme with the pace of my care. There was a man in the room next to me under heavy police escort who was seriously injured and taking up considerable resources (both with his injuries and his behaviour) and a major road traffic accident also came into majors whilst I was there which obviously (and rightfully) slowed my care down further. Then when the surgical team eventually reviewed me they said they needed my blood taking again in four hours so they could see what the extent of my continuing blood loss was and that I should be transferred to a surgical ward as soon as a bed became available. The surgeon reviewing me also told the nurses to get a move on with my fluids....

When I was transferred to the ward I saw one of the surgeons I'd seen as an emergency on a previous admission and he remembered me well. He was great - really efficient and pragmatic. He carried out a rigid sigmoidoscopy at my bedside (which was effectively a rigid perspex pole with a light on the end being pushed into my rectum so it could be visualised) in order rule out any bleeding from my anastomosis. The surgeon was reassured because he couldn't see any immediately apparent reason for the bleed, but he also explained that it needed to be closely monitored.

I told him that my preference would be to be transferred to my local private hospital under my normal consultant surgeon's care in the morning and he said that provided nothing got any worse overnight he was happy with this plan.

But when my second lot of blood results came back they showed that I'd lost a gram of haemoglobin in four hours and the registrar who came to see me with the results explained that whilst the IV fluids may have accounted for some of this drop due to dilution, the loss of blood was significant enough to warrant further investigation. And because of this the hospital was only prepared to release me to my local private hospital if they sent an ambulance for me (which I presumed was to ensure that I actually got there and to cover themselves if anything went wrong during my transfer).

Anyway, I eventually got to my local private hospital by ambulance on the Sunday evening and I saw my usual consultant surgeon shortly after my arrival. We discussed the various tests that could be done to investigate the bleeding, but initially he just wanted to keep me on the fluids and see how I felt over the next twenty four hours or so.

Unfortunately I didn't feel any better the following day and was in agony with deep pelvic pain through to my rectum and also some pain in my upper and lower left quadrants. I was getting pain again where I'd had the internal hernia repair the previous October and I told my consultant that it felt as though the internal hernia had returned and that I wanted it investigating. I told him that I noticed the pain first of all when I was exercising as a tweaking sensation but that it had turned into an excruciating deep pelvic pain that I could feel deep in my pelvis through to my vagina and rectum.

In addition to the pelvic pain through to my rectum I was also getting intermittent pain up my rectum which would, at times, be unbearable when I passed any stool. I was feeling incredibly nauseous and was in so much pain that I was requiring regular IV paracetemol and oramorph up to every two hours. My pain was so bad that at times my heart rate would rocket as high as 150bpm from a normal heart rate of around 65bpm. Furthermore, I was really struggling to maintain euvolemia and every time I came off the IV fluids I would become dehydrated even though I was continuing to drink my E-Mix.

I was struggling to eat because of the nausea and what I found when I did eat was that I would get obstructive type symptoms with abdominal distension and pain and so I barely ate for the entire duration of my stay which lasted from 1 to 21 November.

In terms of investigations, all that an x-ray revealed was some distended bowel loops deep in my pelvis but no sign of any mechanical obstruction or reason for the rectal bleed. My consultant therefore ordered more tests which included a PillCam, capsule endoscopy of my small bowel, barium follow through, colonoscopy and endoscopy.

The PillCam was a revelation in terms of the technology. I simply swallowed a vitamin-sized capsule (after a period of fasting) which was equipped with a miniature video camera and light source. Then as the capsule travelled through my entire digestive tract it captured around eighteen images a minute and sent them to a recording device which I wore on a belt around my waste. The gastroenterologist responsible for the test then viewed my entire small bowel by uploading the images from the recording device to his PC.

It was a couple of days before my surgeon came to the ward to run through the results of the PillCam and the barium follow through and when he visited I was in agony. I was alternating between pacing up and down the room in excruciating pain and then crouching at the foot of my bed in a foetal position which momentarily eased the pain....

And when my surgeon entered my room he quipped:

*"Well I've never seen you in that position before"*

and tried to make light of my agony, presumably to see if I'd smile. I made a vague attempt to smile but I was just so fed up of the pain that I actually looked at him with mad eyes as if to say:

*"stop taking the piss"*

and simply asked if he had any bright ideas to get me right.

He said he'd been thinking and would come onto his plan shortly but first he reassured me that the PillCam revealed no abnormalities and that the findings of the barium follow through were consistent with the x-ray in that all that was revealed were some distended bowel loops deep in my pelvis, indicative of adhesions in this area. I managed to mutter that whilst this was good news it was also very frustrating for me because it didn't shed any new light on how to resolve the horrific pain and nausea I was experiencing, nor did it explain the reason why I'd bled.

My surgeon didn't respond to this and simply said that he'd been wondering whether a permanent SNS device would help with my pain and he suggested that I put my tibial nerve stimulation device on whilst he stayed with me for a while and that I should let him know if this had any impact on my pain.

So I got the device, leads and electrode pads out of their pouch, and slowly sat on my bed (still in agony) placing the pads on the inside of my ankles and calves and switching the machine on before lying down to see what happened.

After about ten minutes the nerve stimulation was starting to ease my pain, to the point where I could actually hold a conversation again. My surgeon therefore suggested that we ought to trial a permanent SNS implant as this would act in the same way as the tibial nerve stimulator which was helping with my pain. I agreed that this seemed sensible but I also asked about whether he would divide the adhesions which we suspected were deep in my pelvis at the same time as fitting the implant.

To this he simply stated that he would be very reluctant to operate without any clearly defined *"target"*, particularly because I'd already had so much surgery. He said that he would discuss further adhesiolysis with the gastroenterologist I'd travelled to see earlier that year (and who'd started me on the E-Mix), but that his initial thoughts were that I should trial the SNS first.

The gastroenterologist agreed with the plan and the operation for the temporary SNS implant was scheduled for 20 November.

I vividly remember the night before my surgery because my pain that evening became so intense that I simply couldn't sit, lie or even stand still. It was worse than the previous night and I was again pacing up and down my room and alternating this with squatting down at the bottom of my bed with my arms outstretched holding onto the foot of the bedpost, remaining in this position until the waves of pain built up again, at which point I would return to standing and pace my room.

I was alternating squatting with prowling and the agony was as bad as labour pain. I was literally screaming out with the pain..... I was getting only limited relief with oramorph, and I found that the best pain relief came when the nurses tanked IV paracetemol through my cannula as quickly as possible. My blood pressure, which was ordinarily fairly low, was shooting up to 165 over 120 and my heart rate was exceeding 150bpm due to the agony I was in. Bizarrely my temperature was also slightly raised at 37.9C and I was beginning to wonder how I would get through the night with the pain.

But with the help of a sleeping tablet and the heavy duty pain relief I did manage to get through the night. And by the next morning I literally couldn't wait to get onto the operating table because I knew that once the SNS was up and running it had a chance of getting my pain under control.

The surgery for the temporary SNS device is extremely minimally invasive, and only a tiny cut needed to be made in my lower back for the SNS wire to be inserted into my sacral foramen. So I woke up from the anaesthetic with little or no pain from the operation itself, but I felt like I had some urinary symptoms and my deep pelvic pain was still there. Then by the time I'd come around completely I realised that my pain was still there because the device wasn't yet connected to the wire and when I got up to go to the loo, not only did I feel like I could have a urinary tract infection (UTI) coming on, I realised that I had some bloody vaginal discharge to boot.

I was aware that the sacral nerve innervates various aspects of the pelvic floor and that SNS can be used for urinary incontinence so I didn't dwell on the additional symptoms I was experiencing and just put it down to the consultant having tested the device in theatre and therefore *"waking up"* these nerves.

That night I was again dosed up heavily with pain relief and then early the next morning my consultant arrived with the device to switch the SNS wire on. We also agreed that I should be discharged home that day (which was a Friday) and keep a symptom diary documenting my pain levels and bowel movements following the SNS implant.

In terms of my symptoms, there was no doubt that the SNS took the edge off my agonising pain, but I was still using regular oramorph at home and instinctively knew that there must be something else going on for my pain to be that bad. I therefore returned to see my consultant in clinic on the following Tuesday and said that I thought the SNS was helping with my pain and also that it seemed to be slowing my bowel movements slightly but that I felt that there was still a clear indication for surgery. I was continuing to experience horrific pelvic pain I was also experiencing increased pain on defecation.

My surgeon explained that he'd performed both a colonoscopy and an endoscopy in theatre whilst I was asleep (and before he put the SNS wire in situ) and that from those tests he couldn't see any abnormalities in my gastrointestinal tract. My response to this was that the pain came on in exactly the same way as it did with the internal hernia I'd had repaired the previous October and that both the x-ray and barium follow through had indicated the presence of adhesions deep in the pelvis and that I wanted these dealt with.

My surgeon's retort was that he was fearful that further surgery could do more harm than good and that he'd only be prepared to operate if he had a definitive *"target"*. At the moment he explained that the only indication he had was possible adhesions deep in my pelvis. He said that he'd like me to go back and see my gastroenterologist who first put me on the E-Mix, together with my gastroenterologist's surgical colleague and that we should take things from there.

I said that I didn't want to wait too long to have the permanent implant fitted and that I didn't want to be without the SNS over Christmas because whilst it was not giving me complete relief it was helping to take the edge off the agony. We therefore agreed that I would have the permanent implant fitted on 4 December (which was in nine days' time) and that I needed to see my gastroenterologist and his surgical colleague before then for their opinion and that I could have one of them remove the temporary wire.

I therefore made the long journey to see my gastroenterologist and his surgical colleague on 1 December, just three days before my scheduled date for the permanent implant. Their firm advice was that surgery for adhesiolysis ought to be a last resort and that if, following the SNS implant, I still suffered with pain then perhaps I should consider a stoma because that may relieve some of the deep pelvic and rectal pain as it would mean that stool would bypass this area.

I was horrified at this thought and was totally adamant that at forty years of age, with a figure that still looks relatively good in a bikini I wasn't having a stoma. I was also concerned that because of my high output any stoma would also be high output and the thought of having to change it all the time just filled me with dread.

Between us we therefore decided that the plan should be to continue with the permanent SNS implant on 4 December with a review of my progress in the New Year.

So I returned home for two nights before going back into hospital again for the operation to fit the SNS device. The operation went well and I had a short overnight hospital stay following which I was promptly discharged home again.

But two days after my discharge I developed a bad UTI and Bacterial Vaginosis (BV). I woke up with a stinging sensation in my vagina and intense pain on urination. I was also passing bright red urine, so I went to my GP who confirmed both the UTI and the BV.

The UTI was so strong that my white cell count rose to 36,344/ul (with normal being 0-39) and my red cell count was 48,991/ul (with normal being 0-50) and I didn't respond to two different sets of oral antibiotics.

Neither did my BV respond to oral or vaginal antibiotics. My GP was also concerned that due to my recent PR bleed, extremely high level of white and red blood cells present in my urine sample, which also showed a mixed bacterial growth, that I may have developed a fistula. (Although looking back now I wonder whether I'd started to brew this infection even before the fitting of the temporary device on 20 November because before that operation I was spiking a fever.)

Anyway, I was referred back to my surgeon by my GP for further advice and investigation and by the time my appointment came around I was feeling like death and was still running a temperature. So I was promptly readmitted for the administration of IV antibiotics with known sensitivity to the UTI and first line IV antibiotics for the BV.

During my admission I had an abdominal ultrasound scan performed, as an initial diagnostic test to rule out any fistula and I was in complete agony when the radiologist passed the doppler over my pelvic region, so much so, that I could only bear the lightest of touches in that area. I then had a follow up CT scan which thankfully ruled out any fistula, although the existence of a *"new"* kidney cyst was picked up.

Thankfully, both the UTI and BV resolved during the admission with the IV antibiotics but I still didn't have any answers as to why I'd bled from my gastrointestinal tract and developed such bad a UTI and BV infection shortly afterwards.....And I was still in intense pain.

In terms of the UTI, I started to question whether I was actually starting to suffer with some kidney damage due to the amount of extra salt I consume, but I've been reassured that this is unlikely and despite the fact a further scan has revealed that the cyst has grown in size I'm just trusting this advice for now……..

As for the deep pelvic pain that I'd been experiencing right through to my vagina and rectum since my PR bleed, this continued over the following few weeks, despite the fitting of the permanent SNS device. And whilst there was no question in my mind that the SNS device did help with the pain (because I simply wouldn't have been able to leave hospital without it) I was also convinced that the adhesions deep in my pelvis must have been causing, or at least contributing to, the pain.

I knew my surgeon didn't want to operate again because the exact location of the adhesions was unknown and also because I'd already had so much surgery, so I started to investigate whether there were any non-surgical solutions for adhesions.

Through my research I came across a practice, headquartered in Florida, USA (but with operations throughout the USA and also in the UK) which uses physiotherapy (or *"physical therapy"* as the Americans call it) to break down adhesions. So I made enquiries and had an initial telephone consultation with one of the therapists based in Florida.

The therapist was really knowledgeable and she was convinced that she could help me, but was also cognisant of the fact that further surgery may also be required. She explained that if I did require further surgery then this would need to be carried out prior to any physical therapy and that I'd then need to wait at least ninety days for any such physical therapy. The best advice she gave me, however, was that if I did opt for surgery I should insist that it be done by laparoscopy using heated humidified $CO_2$. She explained that the use of heated humidified $CO_2$ in theatre would reduce the risk of any further adhesions.

I had a really good consultation with the therapist and when she advised that heated humidified $CO_2$ is used frequently in the USA to reduce abdominal and peritoneal trauma I immediately wondered why on earth it wasn't offered here in the UK. No doubt cost would come into it, but I reasoned that this is surely a false economy because even if it costs a little more, the additional cost would actually make more economic sense in the long run if it reduced the risk of further adhesions and thus the risk (and hence the cost) of further repeat surgeries.

So following my consultation I researched the use of heated humidified $CO_2$ (thanks to Google® and the internet......how we ever lived without it I'll never know?!?!) and I noted from my research that clinical trials had been carried out on rats using a control group who had no surgery, a group who had surgery using cold, dry $CO_2$, a group with heated (37°C), dry $CO_2$ and a group with heated (37°C) and 100% humidified $CO_2$.

What the study showed was that the peritoneal samples taken from the rats who had surgery using heated humidified $CO_2$ were comparable to the peritoneal samples taken from the no surgery group, whereas the group who had cold, dry $CO_2$ had the worst evidence of peritoneal trauma and adhesions. The group treated with heated $CO_2$ had peritoneal samples taken which showed less evidence of trauma and adhesions than the cold, dry $CO_2$ group, but were still worse off than the no surgery control group and the group who had surgery using heated, humidified $CO_2$.

Armed with this knowledge I then went back to my surgeon and asked for his advice on whether I should try the physiotherapy (which was going to set me back around eight thousand pounds) or whether I go for a laparoscopy with heated humidified $CO_2$ first (which my health insurance would cover). I made a point of looking him straight in the eyes and long and hard when I asked him this because I wanted an honest and considered answer. He remained quiet for what felt like at least five minutes and so I interrupted the silence with a further question:

*"what would you do if you were me?"*

He remained silent for a little longer and I continued to look him in the eyes because I wanted him to know that I needed his honest answer. Eventually he broke my gaze and said:

*"I would go for the laparoscopy with heated humidified CO2 because the recovery time will be quick and with the heated humidified CO2 there is a lower chance of post-operative adhesions....plus you'll not be out of pocket."*

*"Great"* I said. Before continuing with:

*"I'm sure you'll find adhesions deep in my pelvis and maybe near my cervix and rectum because I've had bleeding from my cervix and the pain is so deep in my pelvis and feels like everything inside is joined up. I also get lots of pain on defecation when my stool thickens up and as the endoscopy has revealed no "internal" reason for this pain (or the bleed) maybe there is a reason on the outside.*

*I also want you to promise me that you'll look at the entire length of my bowel for adhesions too because I've been getting quite a lot of left flank pain which I initially thought was my kidney ......and I really want this surgery to be the last surgery I ever have ......"*

My surgeon duly reassured me that he'd look carefully at the entire length of my bowel and rectum and that he would also ask my gynaecologist to be present so that he could cauterize my cervix which would deal with any remaining inflammation following my recent episode of BV.

But what I really wanted to know was why on earth they didn't routinely use heated humidified CO2 if it's known to reduce the incidence of post-operative adhesions, so I put this to my surgeon.

When I asked this question he simply said *"cost"* whilst simultaneously rubbing his right thumb against the inside of his index and middle fingers….. And whilst I could understand this rationale in the cash strapped NHS system I had difficulty with understanding it for the private sector, particularly because it's a false economy anyway. I was also slightly concerned about what'd happen if the operation had to be converted to a laparotomy, because my surgeon didn't know what he'd find. To this he said that he would obviously have to *"consent me"* for a conversion to a laparotomy and that if this did become necessary he'd just use the machine generating the heated, humidified $CO_2$ to waft it over my abdomen in an effort to keep the area as warm and moist as possible. This, he explained, should help to reduce the chance of post-operative adhesions.

This all seemed to make perfectly good sense, as indeed did the use of heated, humidified $C0_2$ more generally. In fact, if you give the concept some thought, the abdominal cavity is warm and moist and so by keeping it warm and moist during surgery with heated, humidified $CO_2$ there is bound to be less trauma because the environment inside the abdomen is remaining more constant. It's hardly rocket science, yet it's a relatively new idea.

So, following this consultation I was duly booked in for the operation two weeks later. This gave me enough time to go back and see my gynaecologist before the op, but it was also close enough for me to bear the intervening time in pain because the prospect of relief from the agony was firmly in sight.

On the day of the operation itself I was really nervous because I had no idea what they might find and I was dreading the prospect of the surgery having to be converted to a laparotomy. But I was also anxious to get down to theatre as quickly as possible and to get the op over and done with.

I remember coming round from the operation very distinctly. As I was regaining consciousness I realised I was in the recovery room and looked up to see a recovery nurse speaking with my anaesthetist. Whilst still half asleep I'd obviously been asking the recovery nurse what happened in theatre, because I could hear the nurse saying to the anaesthetist:

*"she keeps asking what was found, how long the operation took and where the adhesions were……"*

So, the anaesthetist then came over to me, squeezed my hand and said that I was in theatre for less than two hours and that my surgeon and gynaecologist were really pleased because:

*"they hardly had to do anything. They divided one band of adhesions, cauterized your cervix and that was about it".*

Great! I thought to myself … but that really doesn't tell me a bloody thing! But before I had time to react and ask any more questions she was gone.

Then as I became more and more conscious I couldn't believe how little pain I was feeling and by the time I got back to the ward I could easily transfer myself from the theatre trolley to my bed and shortly afterwards I could walk easily from my bed to the bathroom with barely any pain.

By that evening all I required was a bit of paracetamol. And the agonising shoulder pain I'd felt following previous laparoscopies was absent as was the post-operative bloating….. Clearly the heated, humidified $CO_2$ was being readily absorbed by my body and the pain from dividing whatever adhesions they'd found was less than the pain that the adhesions were actually causing me in the first place! Hurrah!!

When my surgeon came to see me the next day I was so excited to tell him how great I felt after the operation, but I was also mindful that I didn't want to sound too smug because the push for the operation and for using the heated, humidified $CO_2$ came from me!!

I was also keen to know what they found to see whether, as before, I'd been right about the location of the adhesions...... and when my surgeon confirmed that there were two bands of adhesions, exactly where my most troublesome symptoms were I smiled to myself, because once again I was right.

My surgeon described a *"curtain"* of adhesions joining my rectum to my cervix, which he divided, and a further adhesion joining together a loop of small bowel in my upper left quadrant that he left in situ. In terms of the adhesion in my upper left quadrant he explained that in order to divide this he would have had to convert the operation to a laparotomy and that the therapeutic benefit of doing so was outweighed by the risks involved.

I certainly appreciated that a considered approach had been taken by my surgeon but I was also a bit miffed that he'd left an adhesion in situ because I really wanted this operation to be my last ever surgery. My surgeon explained that the adhesion adjoining the loop of bowel together was unlikely to be giving me any symptoms because the bowel was not *"out of place"* in any way with this adhesion, it was just joining the loops of bowel together where it would ordinarily be lying next to each other. And whilst I got his rationale, he wasn't the one who was admitted to hospital a few months earlier with agonising left flank pain. And even as I am writing this now I still get intermittent flank pain which could potentially be due to this adhesion. I've had fleeting thoughts about more surgery to divide this adhesion, particularly because my surgeon knows exactly where it is and so any surgery could be done via laparoscopy and with heated humidified $CO_2$.... But I really want to stay out of operating theatres if at all possible. And just to complicate things a little (yes, as will be evident by now, things are rarely straightforward for me where medicine is involved) I've also developed this cyst in my left kidney, which I'm now having monitored via ultrasound just in case this is the source of my pain.

And whilst I've been told that the cyst is very unlikely to be problematic, at the back of my mind I do worry about kidney and heart disease because of the amount of salt I now take in my diet. In addition to having the salt in my E-Mix I can also crave salt in my food and have been encouraged to add salt to my food if I crave it because it's my body telling me that my sodium is on its way down.

Anyway, following my discharge from my last surgery I needed another IV top up about five weeks later and a further top up just short of eight weeks after that. But what I'd started to notice following the SNS implant and my last surgery is that when I was drinking a full litre of the E-Mix all the time I was starting to get some water retention in my legs and also my blood pressure increased slightly. I always used to sit around 110/70 or 100/60, even when well hydrated but during my last admission, once I'd stabilised my blood pressure was around 120/80 or even slightly higher on the systolic, so whilst not hypertensive or "*high*" in medical terms, the increased salt intake seemed to be starting to affect me. Yet despite getting some fluid retention, if I dropped my E-Mix to 750 or 500 millilitres for any length of time my sodium levels started to drop quite quickly and so I then had to adjust the intake back up to a litre again (or even more) so as to try and avoid having to be hospitalised. And then when I did this I started to retain more fluid again in my tissues.

The only thing that I was aware had changed in terms of my physiology for my E-Mix requirements to have reduced slightly was that I'd had the SNS implant fitted and I was starting to wonder whether this could make my gut absorb more sodium, but equally didn't understand how.

So I went back to see my gastroenterologist to discuss whether I should reduce my E-Mix and then just increase it again if my urinary sodium fell below 20mmol/L in an effort to stay out of hospital. I also wanted to know whether there were any downsides to taking so much salt and having to keep getting hospitalised for IV fluids, his opinion on whether the kidney cyst could be causing my left flank pain and whether the SNS could be having an impact on my slightly reduced requirement for E-Mix.

As ever, it was worth making the long journey to see my gastroenterologist because he is so good at explaining everything to me and he was completely up front with me if he didn't have all the answers. I think that because he's recognised as being a true leader in his field he's not scared of saying *"I don't know"* when he really doesn't know the answer to something. From a patient's perspective it's so refreshing to hear a medic say *"I don't know"* rather than trying to fudge the answer so as not to appear unknowledgeable or, even worse, dismissing symptoms as psychological because there's no easy answer.

What I've found on my journey is that the human body is so very complex and modern medicine and medics really don't have all the answers. There's so much that's still not known, but most doctors are too scared to admit this, presumably for fear of coming across as incompetent.

Anyway, my gastroenterologist's view was that it's not possible to be intravascularly depleted whilst at the same time retaining fluid in one's tissues and that if I was retaining fluid from time to time then this would be because the amount of E-Mix I was consuming at that time was slightly high for my requirements. But he was also very clear in saying that there's no easy gauge for me and that my daily requirements may fluctuate from time to time.

He advised me that I'm probably difficult to cannulate when I'm admitted not necessarily because of intravascular depletion but because over time as veins get damaged from IV therapy it's necessary to go deeper into the tissue to find veins. He explained that some people eventually run out of good veins and have to have a permanent line in situ.

And whilst I was encouraged that my gastroenterologist thought that this was unlikely to happen to me because I managed to control my fluid status quite well with the E-Mix, he stressed that it's best to try and manage my hydration with the E-Mix alone if at all possible, so drinking slightly less than a litre (say 750 millilitres) when I feel that I have some fluid retention but then increasing it again if my urinary sodium drops below 20mmol/L. But equally, that if my urinary sodium becomes low and my dehydration symptoms acute and I can't then increase my urinary sodium levels quickly by increasing my E-Mix alone then hospitalization for IV fluids would still have to occur. He also said that it's not a good idea to routinely drink more than a litre of the E-Mix a day.

My gastroenterologist also advised that high blood pressure was the main risk of too much salt intake but that if I was taking the right amount of salt via E-Mix and in my food to maintain homeostasis, then even though this may be far more than the regular man on the street may consume, it won't necessarily be damaging for me.

In terms of whether the SNS was helping me retain more fluid, my gastroenterologist was not entirely sure, but did comment that if my bowel movements were slightly less frequent following the SNS implant then this was a possibility. I then questioned whether the SNS could have some effect on the gut's ability to absorb better and to this he said he didn't know. He explained that the way in which sodium is absorbed by the gut and why drinking the E-Mix is helpful is due to the osmotic gradient but because I have my full length of small bowel it's unclear why my losses are so great. He said that there is more than likely a reasonable physiological answer for this and perhaps my ileum is shorter than the "*norm*", or perhaps I have always had an underlying problem with absorption in my small intestine but that it only came to the fore after my colectomy. In short he doesn't know what the reason is yet but he was equally sure that there is a reasonable, physiological explaination.

For me it seems quite straightforward. I still can't help thinking that having my guts out on the operating table during my second caesarean section for such a length of time that they turned blue/grey in colour, must have caused irreparable damage. And given that my ileum now seems unable to absorb the requisite amount of electrolytes, it seems entirely plausible that my ileum is damaged in some way – perhaps in the way I have already described where the villi are shorter than normal due to ischaemia and that the net absorption of water and sodium is impaired.

I've also been told by a natural practices therapist that the fluid retention in my tissues could actually be due to my lymphatic system being under stress and not working as efficiently as possible due to my chronic dehydration, rather than E-Mix overload and following a lymphatic drainage massage any retention does seem to reduce somewhat. I've also been told that skin brushing will help with the retention by stimulating my lymphatic system and so I've brought a skin brush and now regularly use it.

But the lymphatic drainage massages don't come cheap and it's so hard for me to tell whether the retention lessening is as a direct consequence of the massages when I have them or due to my fluid intake.

As for the kidney cyst, my gastroenterologist thought that it was highly unlikely to be giving me my left flank pain but agreed that it might be an idea to have it monitored for *"peace of mind"*. And as I said earlier, my last scan revealed that it's grown and also confirmed that I have pain in this area because when the ultrasound doppler was placed over my left kidney I had pain to such a degree that the radiologist suggested that a kidney infection was the most probable cause of the pain. Urine microscopy ruled out any infection on that occasion, but I will continue to keep an eye on my kidney function. This is particularly so because proper hydration is so fundamental to health, impacting on so many areas of one's body. (After all, at least sixty percent of the average adult's body weight is water.....)

As regards the bigger picture, I do get very frustrated at having to constantly manage my fluid balance with the fluid restriction, E-Mix and IV top ups and I get really down sometimes when I'm feeling particularly tired and lacking in energy due to my body battling to maintain homeostasis. But because I'm able to monitor my hydration status using a urine sodium test I'm at least able to lead an almost normal and carefree life, including travelling abroad for holidays.

This is in stark contrast to the first year following my colectomy, when I was too nervous to leave the country (in case I had another sub-acute small bowel obstruction, or collapsed with dehydration and found myself without access to my usual medical team).

And now that I know the signs of becoming dry I'm pretty good at managing myself. When I'm starting to *"dry out"* I can often feel fluid being drawn from my tissues like a *"glug, glug, glug"* feeling as if my veins are literally syringing out the fluid from the tissues ..... At first I thought this feeling was fat melting away (because in my early recovery period my appetite was poor) but I've since come to recognise this as fluid moving from my tissues, presumably to my veins and this happens most markedly shortly before my sodium levels drop very low. (So perhaps the retention is also my body's own evolutionary mechanism of trying to cope with chronic dehydration, in much the same way that camels humps store their water? Who knows?!)

What I've also found, though, is that I can deteriorate quite quickly once this sensation starts and from then on I'm generally on a downward spiral which can be difficult for me to control without admission for IV fluids. So when my symptoms get acute, particularly with postural hypotension and severe nausea, together with the low urinary sodium and the *"glug glug glug"* feeling I tend not to mess around and get myself admitted as soon as I can. But my longest stint in between IV top ups is now three months and I'm hoping that in time I might be able to manage even longer, and eventually maybe even manage without.

I've learned through this process thought that I must prioritise my health, not only for myself, but also for my family, and in truth, it's being well for the sake of my husband and my children that keeps me on the straight and narrow. And whilst having to drink the E-Mix, carefully monitor my fluid intake and be hospitalised overnight every so often for IV fluids isn't ideal, if that's all it takes to keep me well then it's not so bad after all. There are millions and millions of people in far worse situations than me and in many ways I'm very blessed.

My problem, after all, is just physiological.

# Dismissive doctors

So my health problems have primarily been physiological i.e. functional as opposed to pathological i.e. disease, or even psychological.... And the care I've received and continue to receive from the medics whom I trust and who have really helped me through my journey to my "*altered normality*" has, on the whole, been fantastic. They've listened to me, and not dismissed my symptoms as psychosomatic or insignificant, and they have enabled me to now live a more normal life.

This is in such stark contrast to some of the other medics I've encountered along my journey. Those that haven't listened and have failed to come up with any diagnosis or effective treatment plan have generally had a readiness to dismiss my physiological problems as psychological and have suffered from a "*God complex*". And whilst I'm very ready to admit that psychological or mental health problems, or even extreme stress can have a physiological effect, such as a racing heart, and diarrhoea before an exam, or an inability to think straight or concentrate whilst in the midst of an anxiety attack, I also know that I can tell the difference.

My poor physiological health has had a psychological impact (and in the first year following my colectomy I suffered with anxiety attacks triggered by flash backs to when I was acutely ill with a significantly raised CRP) but equally, when something is physiological, which my adenomyosis, rectal prolapse, redundant colon, anastomotic stricture and inability to maintain euvolemia have all been, I know it; and believe me, if any of these things could have been psychological instead then I'd have taken that in exchange, any day of the week. At least psychological problems can be treated effectively without surgical intervention, and often without hospitalization. But far, far too often I encountered doctors who were ready to dismiss my physiological problems as psychological, primarily because they lacked the skill and/or the knowledge to diagnose me properly.

In terms of what I mean by a *"God complex"*, I've borrowed Wikipedia's current definition (at the time of writing) which is:

*"an unshakable belief characterised by consistently inflated feelings of personal ability, privilege, or infallibility. A person with a God complex may refuse to admit the possibility of their error or failure, even in the face of complex or intractable problems or difficult or impossible tasks, or may regard their personal opinions as unquestionably correct."*

So whilst Wikipedia certainly isn't the fountain of all knowledge, this definition is pretty spot on; and unfortunately too many doctors I've met have been afflicted with a God Complex fitting this definition. They are invariably the ones that haven't listened, or have cut me short midway through my history and have instead stood at the bottom of my bed (not having the time to sit down and engage in conversation) and have talked at me. They are the same ones, who when they can't come up with a diagnosis, have intimated that the symptoms have all been in my head. These doctors simply can't accept that they either don't know the answers or can't find the cause of a patient's ailments and so their *"inflated feelings about their ability and their unshakable belief in themselves"* leads them to tell the patient that they are either psychologically or psychiatrically disturbed.

A particular favourite I've come across is to tell the patient that their symptoms are entirely consistent with a panic attack.

Of all such encounters I've experienced (and there have been many), there are a few that really stick in my mind. An early episode which is particularly entrenched is when I was repeatedly hospitalized the year before I was diagnosed with a redundant colon, slow transit constipation and rectal prolapse, when a Chinese registrar on his ward round simply dismissed my agonising abdominal pain as trivial. Without even taking the time to properly examine me he declared that there was absolutely no need for me to be in hospital and that my symptoms were all in my head. Not only did this anger me, it also delayed my proper diagnosis.

The second episode which really stands out was when I was back in hospital with obstructive symptoms shortly after my second major open surgery. By this point I'd already had four endoscopic balloon dilatations to stretch my anastomosis and a recent surgery to repair an internal hernia, so as you can imagine, when I felt the same agonising abdominal pain and bloating (so familiar to me by this point of a sub-acute intestinal obstruction) I was absolutely gutted. I was admitted via A+E to my surgeon's NHS Hospital where a professor of surgery dismissed me in a bedside consultation (that lasted no longer than about ten minutes). This professor couldn't come up with any cogent reason for my repeated sub-acute intestinal obstructions other than *"things just settling down"* and because he was unable to determine a cause he simply said he was happy for me to go home and suggested I take up yoga or pilates or practice meditation - with the clear inference being that I was over anxious about my health.

This really bothered me at the time because I was so far away, health wise, from being able to take up yoga or pilates that I knew he wasn't even taking my symptoms seriously (despite dilated bowel loops clearly showing on my x-ray). Rather, he was simply questioning my mental health and suggesting coping mechanisms to address any underlying anxiety. It was only many months later when my current gastroenterologist explained that the obstructive episodes after the dilatations and hernia repair were almost certainly a result of acute dehydration that I breathed a sigh of relief (partly because there was no other functional cause, but also because it now all made perfect sense).

My third real encounter with a dismissive doctor with a *"God complex"* was shortly after I'd been transferred to my local private hospital following the comments of the professor on the NHS ward.

As my pain had started to ease and the obstructive symptoms pass with several litres of IV fluids it was clear no surgical intervention was necessary and so my surgeon referred me to a gastroenterologist colleague for his input. This was the same gastroenterologist who (unlike my current gastroenterologist) had quite spectacularly failed to identify the cause of my nausea and small bowel obstructions as acute dehydration earlier that summer. This time, he told me that he thought that I had just been through a lot and that some of my symptoms were undoubtedly psychological and that I should:

*"go home and enjoy my family".*

Again this really annoyed me because the implication was that I was well enough to do this. Clearly if I felt well enough to be at home with my family I would have been, rather than lying in a hospital bed on my own. After the amount of time I'd spent in hospital already that year, being in that room feeling like death warmed up was the last place I wanted to be....... (I also strongly suspected that because he too worked with the professor who'd told me to do meditation, that between them they'd decided I had psychological or psychiatric issues that needed to be dealt with.)

Anyway I informed him that if I felt well enough I would be asking to be discharged and that I was trying so hard to force myself to get up and walk around and to eat to try and see if I would manage at home but that I just felt so dizzy, nauseous, weak and lethargic. I also remember asking him if there was anything else I could do to try and help myself. To this he quipped that I should stop taking paracetamol and antiemetics, IV. Well, this just infuriated me even more because the only things that would stop my horrific nausea were the IV antiemetics and the best pain relief was IV paracetamol, but his clear implication was that by continuing to request IV medication I was in some way setting myself back. So I told him that I would try and take my medication orally again to see how I fared but I was really struggling to cope with the pain and intense nausea.

I then had a flash of inspiration and recalled that my nausea had previously been at its worst when my serum potassium was low, so I pressed my call bell for the nurse and asked if she would take my blood to see if my potassium had dropped. The senior nurse on the ward answered my bell and took an arterial blood sample which could be tested immediately (for various things including sodium and potassium) and sure enough my potassium was low. So despite the IV fluids my electrolytes were still out of balance which explained my horrific nausea. But rather than bother to come and see me again and offer any sensible advice on how to manage my hydration at home, the gastroenterologist simply prescribed some more IV fluids containing potassium and then sent me on my way. And it wasn't until six months later when I travelled to meet my now current gastroenterologist that I was put on a more structured regime to keep my fluid balance in check.

The next time the *"mental health card"* was pulled out on me was during a visit to A+E, where I'd been taken by ambulance because I was too weak, nauseous and light headed to drive myself there. I was so dehydrated that my muscles had started to cramp and when I walked I felt like that puppet on strings. I barely had the coordination required to call the ambulance.

For whatever reason, I didn't hit the required threshold to be given IV fluids by the ambulance, but on arrival I was handed over to the Consultant in charge who immediately recognised my acutely dehydrated state and prescribed IV fluids and antiemetics. Yet when I was in my cubicle about to be cannulated, as I was sipping some squash from a juice bottle to try and combat my dry mouth, a junior A+E doctor questioned the Consultant's judgment about the prescription of IV fluids saying to the consultant:

*"but she's drinking"*

and when the Consultant had left, proceeded to tell me that she was only giving me the fluids and antiemetics because her consultant had said so but that I was almost certainly just having a panic attack because of everything I'd been through and that's why my heart was racing and I felt so strange…….

I was too weak to bother to try and tell her otherwise, and the reality of the situation was that my heart was beating so fast, not because I was having a panic attack but to try and keep my blood pressure up, which was dropping dangerously low due to dehydration.

Without doubt, however, the most extreme case of a dismissive doctor with a *"God complex"* I have encountered was the physician who thought it was a good idea to have me take forty eight slow sodium tablets and drink at least four litres of fluids a day. He didn't listen properly to any of my concerns about the (wrong) treatment plan he'd put me on and instead told me not to be anxious when I was near collapse, fighting to stay conscious in my hospital room because he wouldn't put me on IV fluids when I was acutely dehydrated. He also told me I ought to see a psychiatrist or psychologist because he felt that at least some of my symptoms were *"beyond somatic"*.

So I consulted a psychiatrist following my discharge from hospital and (needless to say given his job) he was a very good listener. His assessment of me was that I was exhibiting no obvious signs of anxiety or depression and that I was coping remarkably well given the nature of the prolonged ill health from which I was suffering. He said that it was a testament to my strength of character that I was holding up so well and that he didn't think that my medication needed increasing. (I was already taking duloxetine and amitriptyline which help both neurological pain and stabilise mood as I'd started to take these meds shortly after I began suffering with the anxiety attacks triggered by flash backs to when I was critically ill.) He did, however, say that I could possibly benefit from seeing a psychologist given the adjustments I had to make to my lifestyle due to my ill health and that he'd be happy to refer me.

I was unconvinced about the benefits psychology could bring but heeded my psychiatrist's advice and went to see the clinical psychologist to whom I'd been referred. The lady psychologist I saw was lovely and also very bright. She quickly identified four areas that I could work on to help improve my overall sense of wellbeing, but she was also quick to remark that I seemed to be coping exceptionally well with my *"altered normality"* and that it was up to me whether I wanted to take the sessions further. The areas she identified were:

1. *my concerns about my children's insecurities/anxieties about my health;*
2. *my anger at the loss of opportunity with my career;*
3. *my feelings of unfairness about what has happened; and*
4. *my constant frustration at being unable to match my activities to my energy levels.*

In the end I only went to the psychologist that once, because I felt that identifying the areas I needed to work on was therapeutic in itself and I also felt that because these areas had been identified I could work on them alone if I felt the need.

I've not ruled out going for more psychological therapy in the future – but at the moment I feel mentally very strong and excited about my life again. What I have, however, discovered from my experiences recounted above and from speaking with others who have had complex health issues, is that when some doctors can't easily diagnose or treat you they write your ailments off as a mental health issue. And this is particularly the case for those doctors who suffer from a *"God complex"* because they clearly believe that if you were truly physically ill they would know what to do.

For a patient it is really confusing and frustrating to be told that there is no reason for your physical ailments, because when you're poorly and the cause can't easily be found, you invariably have a certain level of anxiety about the status of your health. So, when these doctors inform you that your illness is more likely to be psychological, you start to question your own mind which increases the anxiety further and a vicious cycle is created. And whilst I have no doubt that in some cases patients may well be suffering with mental health problems in addition to, or rather than physical ailments, my experience is that too many doctors dismiss complex cases as psychiatric or psychological in aetiology without exhausting all investigative avenues. This also means that doctors are failing to diagnose real physical health issues with alarming regularity.

If only these medics listened more.....

If they did, they would be able to get their patients back to health, rather than leaving their patients poorly and with no alternative but to search for answers elsewhere.

# Holistic and inclusive medicine

So when one's told by doctors that the problem is psychological or that there is nothing further that can be done it is often to alternative medicine that people turn. And whilst some alternative practices are widely questioned by allopathic medicine I have also discovered that there is a real place for alternative or more "*holistic*" practices in our healthcare system. Unfortunately, however, the medical profession in the UK does not, on the whole, embrace these concepts.

Towards the end of my journey, and because I was so convinced that I was starting to retain fluid in my tissues in between episodes of dehydration I decided to explore whether bio-resonance testing and therapy could help me. (I had a couple of friends who were experimenting with this around the same time and who had independently of each other recommended that I try it.) In particular, I instinctively knew that there must have been a reason why I was starting to retain fluid, whilst at the same time getting episodes of dehydration (confirmed with a low urinary sodium) and wanted to see what bio-resonance had to offer.

Bio-resonance testing and therapy works on the principle that every substance is composed of matter originating at the atomic level and atoms (comprising protons, neutrons and electrons) have their own net electric charge. So, in our bodies, every tissue and organ has a unique electrical frequency, as do toxins, viruses, bacteria and parasites, and as such, when our bodies are exposed to physical, emotional or environmental stresses, the first thing that happens is that the frequency of the tissue or organ changes. Thus with a bio-energetic screen any vitamin or other nutritional deficiency can be assessed as can any allergies, immune system deficiencies, parasites, viruses or bacteria affecting the body *and crucially*, where in the body the major electrical disturbances are located.

Deficiencies are then treated by taking naturopathic supplements and the areas of the body where the major electrical disturbances are located are treated with the application of electrical frequency therapy. Various different machines are used for such therapy and they work on the basis that the machine collets electromagnetic information from a patient, then normalizes any abnormal frequency information before sending it back to normalize the affected area.

Now, I'm no new age hippy and certainly no bio-physicist (and am still intrinsically cautious and more than a little sceptical about alternative medicine having been raised in a traditional medical household), but the underlying principle that we are all just made up of atoms and that all atoms have an electrical frequency is very basic science and something that I knew made sense. Moreover, I was absolutely stunned with the results of the vitamin and mineral deficiency testing together with the results of the *"problem areas"* of my body.

The bio-resonance screening revealed that I was sodium, magnesium, Vitamin E, Vitamin C and Vitamin B2 deficient and these findings correlated with what I knew about my health from blood and urine tests carried out at various points in the preceding twenty four months.

As for the areas of my body that needed bio-energy treatment, the machine came up with my gallbladder, intestines, left kidney, bladder, lymphatic system, thyroid gland, pituitary gland and my back (and specifically my spine at C1, C7, T2 and my sacral area). It also picked up *"compartment syndrome"*.

Well, I need hardly confirm that my intestines require assistance and the other areas identified all resonated to a greater or lesser extent with what I knew from traditional medicine from various times in my life.

So, going through the list in no particular order: I'd had pain in my left kidney (plus the cyst), a recent severe UTI (bladder) and as

I'd been retaining water in my tissues it didn't surprise me at all that my lymphatic system was flagged up as an issue. I've also read that the pituitary and thyroid glands are crucial in maintaining fluid homeostasis, so again these areas being lacking in energy or having altered energy frequencies is no surprise given my chronic dehydration. I've also had intermittent pain in my upper right quadrant where my gallbladder is located (but haven't even wanted to entertain the prospect that I could have any problems there (!) so have largely ignored it). Finally, I know I have a small element of bilateral facetal degeneration in my sacral area because this was picked up when I had an MRI scan of my lumbar spine and I get neck, shoulder and arm pain when writing or typing for any length of time due to right sided thoracic outlet syndrome (a type of compartment syndrome) which was diagnosed as a teenager and which flares up and requires physiotherapy from time to time. My physiotherapist has also confirmed degeneration at C1, C6/C7 and T7 and has since worked on these areas and given me some relief.

The fact that the bio-energetic screening picked up on all of these areas (and perhaps more tellingly didn't pick up on any health issues that I don't have) made me tempted to find out whether bio-energetic therapy sessions could help with the fluid retention and I had two sessions, following which I felt quite energised, but didn't really notice any dissipation of the fluid. And as the sessions came with a hefty cost (and because I'm still a touch sceptical) I decided that I wasn't prepared to part with any more money for further sessions just yet. Instead, I chose to try some lymphatic drainage massages on the back of the identification of my lymphatic system being under stress and physiotherapy for my back.

In my first lymphatic drainage massage my therapist confirmed that my lymph nodes behind my knees, in particular, were blocked and following this massage I had some prompt dispersal of some of the fluid from my tissues. The natural practices therapist who carried out the lymphatic drainage massage also explained that chronic dehydration can, in fact, lead to the lymph system becoming blocked and that she could feel the fluid being retained in my tissues. I have since chatted the concept of dehydration leading to a blocked lymphatic system over with both my GP and my surgeon and they have said, quite independently of each other that this all makes sense.

Now if money really were no object and if I was still feeling really poorly (as opposed to simply suffering from dehydration every so often) I would definitely have gone for the ten sessions of bio-resonance therapy. It is also worth mentioning that the bio-resonance practitioner I saw suggested that if I had therapy I shouldn't need to have my SNS turned on at all moving forward. He even went as far as to suggest that had I seen him earlier I may well not have needed my colectomy or hysterectomy and was also deeply sceptical of pharmaceutical companies and their role in modern medicine. And whilst I don't agree with his comments on my surgeries (because they were performed largely due to physiological, functional problems) I do have some sympathy with negative views on some of the large pharmaceutical companies' practices.

My sympathies stemmed from a particular experience I'd had as a young lawyer in London, not long after the introduction of the Data Protection Act 1998 (**DPA**), whilst I was working on a corporate deal. The transaction in question involved a private equity company looking to take a controlling stake in a healthcare services company. The business model of the target company was essentially a service provided to GP surgeries which would result in a change to their prescribing practices which would benefit the sponsoring pharmaceutical company and the patient.

The target company would send in nurses to review patient records in GP surgeries and look at the drugs prescribed to the patients. The nurses would then suggest alternative drugs to be prescribed, which were deemed as *"better"*.

Without divulging all the detail, my assessment of the legal position was that the practice was unlawful under the DPA. So, my due diligence report, prepared for the benefit of my firm's private equity client, scuppered the deal (and indeed the entire business model of the company thereby wiping out its entire value unless an alternative model were to be implemented). As such the rather nonplussed management of the target company and his advisors persuaded the private equity house that they go to the expense of obtaining the opinion of the leading QC on the DPA before abandoning the deal. So I drafted up the instructions to counsel and together with my supervising partner, two senior representatives from the private equity house and the MD from the target company and his lawyer, we pitched up at the QC's chambers.

I remember sitting around that conference table with bated breath, hoping beyond hope that this eminent barrister would agree with my advice, but at the same time wanting to disappear into the ground because of the look of disgust with which the MD had fixated on me. Clearly he would have made a lot of money from the deal, but the QC confirmed that it was illegal under the DPA and I had stuck my neck on the line to say so ....... he must have hated my guts....... And all these years later this episode sprung to mind when the bio-resonance therapist was waxing lyrical about the pharmaceutical companies and their *"not so wholesome"* practices.

Anyway, I digress..... Back to holistic therapies.

Well, in addition to exploring bio-resonance (which, incidentally, I understand is used in Russia alongside more traditional medicine) and using physiotherapy and lymphatic drainage massage therapy to disperse any water retention, I have also tried reflexology as a way to activate the natural healing powers of my body (given that I am always battling with dehydration and mindful of the adverse consequences this can have on various aspects of my overall health).

In terms of diet I have substituted Himalayan crystal salt (the health benefits of which are widely publicised) for regular sodium chloride in my E-Mix and I take daily vitamin and mineral supplements in an attempt to maximise my health to the fullest extent possible. But perhaps the most fundamental change and one which I think has helped to keep me out of hospital for longer periods in between IV top us is that I now drink Kangen® water. Kangen® water is water which is treated by an alkaline water ionization machine. This turns regular tap water into alkaline, antioxidant, ionised water which has many publicised health benefits.

I have also tried a *"hands on"* natural healing therapist who works with the human energy field, which was a really good experience. But as with the bio-resonance therapy, I'm not quite ready (or perhaps just don't feel quite needy enough or wealthy enough right now) to embark upon a course of sessions.

I do, however, firmly believe in the human energy field existing beyond what we can ordinarily, see, feel or touch and am absolutely convinced that there is a role to play for alternative and natural medicine that works with our energy fields alongside traditional allopathic medicine.

I am equally convinced that if traditional allopathic medicine were to embrace holistic concepts with a more open mind, thus integrating them into our healthcare system we would all be in a better place.

# The impact

Needless to say, the impact of my prolonged ill health has been far reaching, and apart from feeling like I've been to hell and back, I'm now left with a lifelong health condition which I need to carefully manage, because if I don't I could slip into hypovolemic shock or a coma.

I have between six and ten bowel movements a day (all diarrhoea), sometimes more and sometimes less (depending on what I eat), and I wake up at least once each night to go to the bathroom. I'm constantly tired as I never have an uninterrupted night's sleep and my body is continually battling to maintain fluid homeostasis. I suffer from various abdominal pain, particularly low pelvic pain, together with the referred back ache, due to pudendal nerve damage and the adhesions from the amount of surgery I've had, and I often have a sore anus from the constant wiping after toilet visits. The pelvic discomfort I suffer with impacts on my sex life with my husband and the inability to achieve a completely satisfactory orgasm is frustrating in the extreme. I get tantalising close to a full orgasm but am unable to reach that amazing place of all consuming relief. My husband assures me that his enjoyment is not diminished but I'm sure he's just being polite.....

I've been advised against taking any high impact exercise (although I'm starting to ignore this as I've been gaining weight) and will never work full time as a lawyer again (unless it's as a sole practitioner) because my health condition, as it currently stands, renders me inherently unreliable (in an employer's eyes). When I need to be hospitalised for IV fluids overnight it's always on an emergency basis, and it takes me out of the workplace for the best part of two days, at extremely short notice.

Whilst for the majority of time I'm very positive about my life (because I have air in my lungs, a wonderful loving family, an excellent network of friends and don't want for any home comforts) I can feel quite down at times if I dwell on my limitations.

In particular, a big frustration and cause of melancholy is matching my activities to my energy levels. Prior to my health issues I was a complete whirlwind of activity, living life to the max, rushing from one thing to the next, churning out work and managing to fit family and social life in on top….. In my early legal career I would work long hours and play hard, both socially and physically. I had an active social calendar as well as keeping fit and playing sport to a decent level. My mind still runs at a million miles an hour but I'm constantly having to rein myself in, stopping myself from taking on too much work, organising too many social events or offering to do too much for friends and family, and in truth I don't think I'll ever adjust fully to my *"disability"*.

I hate using that word but a disability is ultimately what I have.

I'm currently a much more hands on and stay at home mother than I had ever envisioned, working around my children's schedules, rather than the high flying working mum I'd always been. And whilst this undoubtedly has its advantages, the fact that this way of life has been *forced upon me* due to my health constraints *rather than chosen by me* is an issue for me. In fact I'm still trying to find the right balance.

In blunt terms, I've not felt fully well since the birth of my second daughter over eight years ago and I've spent far too much of my girls' early years poorly, either at home or in hospital. In total I've spent the best part of a year in hospital. I've also spent countless weekends at home in bed where my husband has had to entertain the children alone because I've not been well enough to enjoy family time.

So, as you can imagine, my ill health has massively affected my family, my friends, my work and my finances.

So to my husband.

*My husband*

Well, my husband has been truly amazing throughout my illness. He is such a strong, focussed and wonderfully loving man, as well as being jaw droopingly handsome and very successful. I feel so blessed to have him.

But despite his strength, my chronic illness has, at times, taken an unbearable toll on him. I remember one evening, very distinctly, when I was in hospital recovering from my colectomy. I was over the worst of my illness and my CRP was on its way down, although I still had several drains in my abdomen. (By this point I'd been in hospital about four weeks.) Anyway, along with some clean underwear and a few other essentials my husband had brought in with him, he'd brought me a pile of unopened post.

As an aside it's worth mentioning that my husband never opens the post as it's always been my job to deal with the post, manage the bank accounts, washing, shopping and household affairs more generally. (Although he is amazing at pulling his weight with cooking and tidying and in fact he's a bit of neat freak and is known to tidy away ingredients whilst I'm cooking ….and even before I have used them!)

Anyway, amongst the unopened letters was one from a nearby county police constabulary issuing a speeding fine, which didn't fill me with pleasure when I opened it, but equally wasn't a big deal. But I then opened a letter containing a Court Order, against my husband, ordering him to pay almost seven hundred pounds in fines and endorsing a six point penalty on his licence. Basically, what'd happened was that because he hadn't responded to the speeding fine in time he hadn't produced the required evidence and so the case moved quickly to an award of six points plus the hefty fine for failing to provide information to the police. This really pissed me off and although I was feeling weak I had a real go at him, raising my voice as much as I could (given my unhealed abdominal wound and associated pain) to blurt out:

*"I just can't believe how bloody irresponsible you've been ...... you can't afford to lose your licence. You have a business to run and we both need cars."*

My husband just looked at me ashen faced and bewildered with his eyes filling with tears. He said in a quiet but firm voice:

*"Have you any idea? Any idea at all what these last weeks have been like for me? I've been worried sick about you, not knowing if you would even live through this. I've been trying to keep the kids stable, run a business in the toughest economy in recent memory, fitting in overseas travel commitments and then visiting you as often as I can......I've been so stressed and I've been so lonely .......but I've had to be strong for you, for the kids, for the business and the last thing I need is for you to have a go at me about not opening the effing post. In the whole scheme of things not opening the post really doesn't matter......I'm going to go home love......"*

My heart melted and I looked at him, my eyes filling with tears. I just held out my arms and said sorry a thousand times over. I cried so hard that evening in his arms. I cried about being cross, about being poorly, but mostly because I missed him and my beautiful little angels so very much. My husband held me tight in his arms and we felt our love and devotion for each other, as strong as ever.

He truly is an amazing man.

As for the post .... had I been well enough I'd have appealed the six point award ...... but even when I was discharged the following week I wasn't up to doing it and I've not mentioned the hefty fine and six point award again (until now).

I also worry that there must be times when my husband wishes he'd never met me. I'm his second wife and had he not met me there's a chance (albeit remote) that he may have reconciled his differences with his first wife. He would also not have had the terrible stress of my illness and he'd be financially independent if he'd not had to pay out a seven figure divorce settlement! But then again we wouldn't have had our angels and I know that neither of us would change our paths crossing for the world. My husband often jokes that he thought he was marrying a young fit girl, but what he's been given is:

*"fit for the knackers' yard"*.

In my moments of strength I know that he's only joking, but I still somehow manage to feel guilty about putting him through the pain of watching me suffer so much and for so long.

When I think of my husband I just fill with adoration and want to be strong and look after myself so that I can keep myself in the best health possible for him and for the rest of my family.

*So now to my daughters.*

My eldest daughter was in year one at school when I started to get really poorly and she's been quite amazing throughout my illness. At the time of writing she's now in year five and turns ten at the end of the month. My illness has, however, in many respects, made her grow up too quickly. She often mothers her little sister and has a confidence and maturity that belies her years.

Perhaps the most moving memory I have of my first born during my many hospitalisations was when she brought in some homework for me to read as I was recovering from my colectomy. She'd been learning about the Black-Death and the Great Fire of London in school and the children had been tasked with writing a story for homework. My daughter's story, written when she was just seven years old went like this:

*"Once upon a time there were [sic] a little girl called Charlie. Here [sic] Mummy had the plague. She had to be good for here [sic]. She had rings all around here [sic] body. Here [sic] Mummy was very ill. Then suddenly there was a knok a [sic] the door it was Charlie's friend Issabella she came to help here [sic]. She said I will go and get some vinegar and poses. So Charlie wayed and wayed [sic] the she came back. She had the stuf [sic] to try and make here [sic] better. Then the rats came and hided then Layla rusht [sic] in "I will help you I will" she said, "I will try and get the rats away with you! Now wee haf [sic] to get a candull [sic] so wee [sic] can see".*

*Then they got the rats away and then they were happy. "Now we have to get Charlie's Mummy better". But just they went to the ships and ......but there was a nuther knok [sic] at the door. It was Charlie's daddy and sister they were out hunting for the cats and dogs. Then Charlie's Mummy got better. Charlie's friends were happy. But then Abbie berst [sic] in with happinnus [sic]. Then they all had a big party. The end."*

I read the story with tears streaming down my face and told her how incredibly beautiful and clever she was and how proud I was of her. Then I just held her tight to my chest almost breathing in her love, her fear and all the other emotions that must have been swimming around in that gorgeous little mind of hers .........

And now that I'm over the worst of my health issues and just get briefly hospitalised with the dehydration she is so grown up and so pragmatic about things. She'll say to her little sister:

*"don't whinge about Mummy going into hospital, you know it's only for one night and if she doesn't go she'll get poorly and will then be away for longer!"*

or

*"stop crying, you know Mummy can't help being sick, now just be kind and give her a hug!"*

My eldest daughter is growing up so fast and it makes me cry that I've missed out on so many special moments due to my illness ……… But because of my illness I've given up working full time, so at least on the positive side, when I'm well I'm spending much more quality time with her. And she never ceases to amaze me. Not only is she striking to look at, she is a high achiever in all areas. She is a straight A student at school in both effort and achievement and manages to impress in whatever extracurricular activities she turns her hand to. And despite being a bit mean to her little sister on occasion, she is actually very doting, sensitive and loving and the affection she shows when my little one needs it most is very moving.

My eldest will always have a soft spot in my heart as she was the first born and my youngest because she will always be my baby girl…..

*OK. So now my baby girl*

Well, because I was so poorly in the weeks following my second daughter's birth I didn't bond with her as much as I would have liked. I was so weak and frail following the delivery and had to spend some time in hospital without her in her first weeks. And due to my husband's work schedule (he was the CEO of a plc at the time) I had no option but to hire a maternity nurse to care for her when I was unable to. I also had to give up breastfeeding when she was only six weeks old so missed out on some precious bonding time.

Being a second child, I was not as anxious about her every whimper as I was with my first, but it became apparent quite early on that her traumatic birth had affected her. She was quite colicky, had real trouble feeding and suffered from constipation.

My maternity nurse recommended that I take my daughter to see a cranial osteopath, which I duly did and after six weeks of treatment her colic and constipation had largely resolved and she was feeding really well. So I think that the initial trauma she suffered at birth was dealt with very well early on and thankfully she was too young to remember me being poorly directly after her birth and up to and including my hysterectomy. But it's quite interesting that my second daughter is most definitely a *"daddy's girl"* and I am sure that this is a legacy from her first months when her dad, despite his work schedule, was more present for her than I was.

I then started getting poorly with my constipation and recurrent hospitalisations when my little one was only three years old. She is now eight. So really, for as long as she can remember I've been in and out of hospital and for some of this time I was critically ill. During my hospitalisations my baby always seemed to cope quite well and often asked her dad if she could go home half way through a visit because she'd get bored. But last summer she started getting quite withdrawn and very clingy to me. In school she wasn't mixing well with the other children and had developed a real attachment to one other girl, which was fine when they played well together but if they fell out it was upsetting to both girls and their upset was spilling over into learning time.

Then one morning my little girl almost broke my heart. She started crying uncontrollably just after breakfast for no apparent reason. She was shouting:

*"Mummy, Mummy"*

and because I didn't answer her or come running straight away (I was actually in the bathroom on the loo) she got herself in a complete state and threw herself on the floor whilst yelling:

*"you don't love me do you mummy?"*.

Of course I told her I did, but she continued to cry uncontrollably whilst insisting that I didn't. And so I asked her what was wrong, to which she replied through her tears, as she was starting to calm down:

*"I think it might be better if I was dead. It would be better if I hadn't been born. I want to kill myself."*

BOOM.

Her words were like a bullet into my heart.

Obviously I responded with warm words, lots of hugs and love and held her on my knee with my right arm under her head and my left arm under her knees so she lay in my arms like a baby. I cradled her as close to me as possible and reassured her that I loved her more than anything else in the whole world (alongside her sister and my husband of course). I asked her again what was wrong and what was bothering her but she couldn't articulate what was bothering her beyond telling me she wanted to die and just continued to sob.

Then she squeezed me tightly and started kissing me all over my face telling me how beautiful I was and how much she loved me. She remained very clingy for the remainder of that week and started saying that she didn't want to go into school and that she had no friends and that nobody liked her and that she just wanted to be with me all the time. And if she couldn't be with me she wanted to hide away and pretend that she wasn't here anymore.

This continued for a few weeks and she kept saying she wanted to die and she became even more withdrawn. I was so distraught and upset because up until that point she'd always seemed so happy at school, was very popular with the other children and I had no concerns at all about her wellbeing. She was a very low maintenance child compared with my older daughter who'd always been a bit more of a handful and the high maintenance one in the family.....

A good friend of mine who's known both my daughters since birth is a child psychologist and so I reached out to her as I didn't want to *"medicalise"* my daughter's anxiety by taking her to the GP. I asked my girlfriend if she could come over one morning and assess my little one at home as she hadn't wanted to go to school that morning anyway. Fortunately my friend was free and so she popped over.

After spending half an hour or so just chatting and playing with my little girl my friend had managed to elucidate that my daughter thought that my illness was all her fault. My friend also said that she was suffering from separation anxiety and that I just had to reassure her that I was fine and that I loved her and that my illness was not her fault and to spend quality time with her.

I felt so guilty and stupid when my friend told me this. My daughter had obviously overheard me speaking to people about my illness and me saying that my problems dated back to her birth. She had then decided that it was all her fault and it would have been better if she wasn't born. My precious little girl had been carrying this burden around with her and had become more and more withdrawn and upset and didn't speak to me about it. I just cried so hard when this realisation set in ……..

I've now worked really hard on putting her fears to rest and have explained over and over again that my illness isn't her fault and that she's more precious to me than anything else in the world (together with my husband and her sister).

I also took her to see a child psychiatrist because she would still often say:

*"I'm so sorry that you're poorly Mummy, it's all my fault isn't it?"*

and despite my reassurances she couldn't get over her feelings of guilt.

Taking the step to see the child psychiatrist was a really good move as my daughter benefitted greatly from being able to talk about my illness in a safe place whilst being reassured by an independent person that I'll be fine. The psychiatrist suggested that my daughter should meet with my medical team to better understand my health issues. At first my little one was keen to do this but after a couple of sessions with the psychiatrist I started to explain things to her, and in particular how my tummy looked on the inside after my surgery and how her tummy looked, with the help of pictures on the internet and I know this really helped. In fact after me taking the time to explain things to her and with the reassurance of the psychiatrist she decided she didn't want to meet with my medical team after all.

Time is also a great healer and my little girl is slowly but surely becoming more emotionally stable with time. Giving up full time work has enabled me to spend more quality time with her, which has not only helped with her separation anxiety, but also with her application at school. My little one now mixes better with the other children in class and has gone from a B grade student in effort and achievement to an A grade student across the board. And I do think that she now knows my ill health is not her fault. I do, however, wish on occasion that I'd persevered to bring a successful law suit because that would have been the best vindication for her that it absolutely wasn't her fault but the fault of negligent medics.....

Anyway, that is not what I want to be doing with my life .....

And I am so proud of both my girls, though; at how strong they are at coping at home without me when I fall ill..... especially on those occasions when my hospitalisations coincide with my husband being away on business.....but I still feel cheated for both myself and for them in terms of the precious and irreplaceable time we have lost out on together.

*Now to my mum*

My mum is a wonderful lady. She is one of nine children from a strict Roman Catholic family. She is incredibly strong-willed and independent. But when I became very poorly in the January after my colectomy, my father had only been dead for eight months and my mother was in complete denial about the seriousness of my post-operative illness. She didn't visit me for the first two weeks I was in hospital and then when she did first visit, I still had my nasogastric tube in place and I heard her break down crying as soon as she'd left my room. She quite simply couldn't deal with me being so poorly, particularly as I'd given her great strength immediately following my beloved father's death.

In the immediate aftermath of my father's death I had a month or so off work in between jobs and whilst I was recovering from my resection rectopexy surgery. During this time, and despite the fact I was convalescing, I helped my mum with lots of the admin following the loss of a loved one. I used my legal skills to sort out the will and probate and we spent some lovely afternoons together sorting through and cataloguing my father's old coin collection.

Then when I started getting quite poorly again a few months into my new job I didn't share that fact with my mum. So when I did get hospitalised in the January prior to my colectomy it was a real shock for her and I think she was overwhelmed by it. Initially I was really upset that she didn't visit me, but when I rationalised it and realised that it was because she couldn't face the reality, I just felt guilty. I felt guilty because I was responsible for putting her through stress and anxiety when she already had to cope with adjusting to life without her soul mate of fifty years.

Now that I'm over the worst of my ill health and my hospitalisations will just consist of re-hydration every so often I think my mum worries a lot less about me. But she still worries far too much because I'll always be her little girl and being a mother myself I can understand her angst.

My illness has also brought me very close to an old neighbour whose first child was born just days after my eldest daughter. We have bonded because she has also suffered with a series of complex health issues following the birth a child. It was also my neighbour's second child that was the tipping point for her health issues. My neighbour opted for an elective caesarean at a well-known private hospital for her second birth. The consultant who performed the caesarean left placenta inside her which became seriously infected and she has suffered with a string of abdominal problems and recurrent infections ever since. The consultant responsible for the caesarean has subsequently been struck off the medical register, but this is scant comfort for my friend who has suffered terribly with her health since. She's currently been in hospital for the best part of the last year and a half.....

I've also made a good friend in someone I first met at the tea and coffee trolley on the maternity ward after the birth of my second daughter, when we both bemoaned how poorly we felt and what bad experiences we'd had (little did we know what the future had in store). She too has had problems ever since and ended up with a hysterectomy far too early in life..... We often joke together about how we met and how badly we've suffered since.

Seriously though, what I find astonishing, is that in my smallish circle of family and friends I can name five young women (aside from myself) who have suffered with life changing health problems triggered by negligent care in childbirth. And what is even more worrying is that I don't for one minute think that this is an anomaly because every woman I talk to about childbirth in the UK knows several horror stories involving either the mother or baby and in some really distressing cases, both. Something certainly has to change with our maternity services here in the UK because the current situation is completely unacceptable. We have a third world system, in a supposedly first world nation, which is totally broken.

And because it's broken I've been left on many an occasion having to rely on my nanny and the goodwill of friends and family to help me out at short notice when I've been hospitalised. Indeed I feel truly blessed to have such a good nanny and such wonderful friends and family who have given me so much support during my illness – who have offered me help with childcare, cooked meals for my husband and either sent me or visited me with flowers and gifts in hospital .... and always been at the end of the phone. In particular, my eldest niece was a complete godsend when I was in hospital for any length of time as she just moved in and took over my place in the home providing some stability for my girls and my husband..... And my sister called almost daily when I was at my worst in hospital, offering support and advice not only from herself but also from the medics with whom she worked.

I do feel guilty though from time to time because I've had to cancel so many social commitments over the years. I can't help but feel that people must have got incredibly bored at times hearing that I couldn't do this or that because I was back in hospital again or simply felt too wiped out to keep to an arrangement. But I've now learned that I have to be ruthless with my time and my energy so as to save as much as possible for my husband and my children. Whatever is left I give to my extended family and friends and then work has, quite simply, had to come last.

*And so to work ......*

Work is the area of my life that has changed the most as a direct result of my ill health. When I first started getting really poorly with my altered bowel habit I'd just taken a job as a salaried partner at a small boutique law firm with a predominantly retail client base. I was the sole intellectual property partner at the firm and if I was absent there was no one to pick up the work. During my nine months at the firm I was admitted to hospital four times and was absent from the office for almost seven weeks. And whilst I worked as best as I could from my hospital bed or at home, I couldn't service my clients properly and with the firm being small (and with us still being in the midst of the *"great recession"*) it couldn't afford to carry an injured party.

So immediately following my return to work after my fourth hospital admission in six months (during which I had had my appendix, right ovary and right fallopian tube removed, adhesions divided and some endometriosis resected) the Managing Partner and another Equity Partner called a meeting with me. I was totally shell shocked during the meeting because I'd dragged myself into work barely a week after the surgery, determined to show my work ethic and commitment and instead I was told that they were serving me with three months' notice to terminate my contract. And as I'd only been with the firm for just under nine months I had no employment rights and couldn't really challenge the decision.

In any event as I'd been so poorly and they offered to put me on gardening leave I was almost relieved by the decision. It meant I could take the next three months off whilst being paid in full and without any stress.

Then following my gardening leave I felt much stronger and the operation to remove my appendix, right ovary and right fallopian tube appeared to work (in that my pain subsided for a while), so when the opportunity arose I decided to return to work again as a fee earner at a large law firm where I'd previously worked, with a promise of partnership. Within a few months I realised that I'd made a huge mistake (because the promise was turning into an empty one) and when an old friend called me one day to say she was actively recruiting in her legal team at a household name FTSE 100 plc I didn't take much persuading to attend an interview.

But in between accepting the job and handing in my notice at my law firm I became poorly again and started to question whether I ought to continue with the job move. My hope was obviously that the resection rectopexy operation would sort me out for good and that I would start my next role a *"new woman"*, but I was also wary about moving when my health hung in the balance. I discussed the job move at length with my husband and my father (who was still alive at this point) and because the job was with such a high profile plc and the work mix was going to be so exciting, they told me I'd be mad to turn it down – and they were right.

In terms of timing, the operation for my resection rectopexy took place at the start of the last month of my notice period and I had planned to move straight to the new role so I didn't miss out financially. Fortunately, by the end of the last month of my notice period I was fully recovered and so I started my new job on time. The plan for my new job was that I was to work for six months getting to know the legal team and the business, after which point I would be promoted, take on significantly more responsibility, manage my own team and in turn would be rewarded with a significantly better package.

Unfortunately, however, it became clear quite quickly over that first summer in my new role that the resection rectopexy hadn't been as successful as I'd hoped and that my health was affecting my ability to work. Indeed just before my promotion was to be announced I approached my line manager to tell him that I wasn't up to taking on the bigger role due to my health. Instead of taking the promotion I submitted a part time working request to reduce my days to four instead of five. My line manager accepted my proposal on the basis that as soon as I felt fit enough I'd be expected to return to work five days a week.

I didn't ever return to five days a week. And about a month after reducing to four days I had my colectomy.

I took a total of three continuous months' off work for my colectomy and my employer was incredibly accommodating at first.

I had a phased return to work which was organised through the company occupational health department and I was supported with counselling paid for by my employer when I suffered with flash backs and post-traumatic stress. But because of my continuing problems, first with the anastomotic stricture, then the internal hernia followed by the long hospitalisation for dehydration I only managed to return to work for short periods of time in between hospitalisations. And when I did return to work I was only working two days a week in the office and two days a week at home. After a year had elapsed since my colectomy it had become clear to my employer that I was too weak to return to working four days in the office any time soon; so I was eventually persuaded to come to an agreement to terminate my employment.

Ultimately my line manager had put his foot down because he couldn't deal with the inherent uncertainty of when I would fall ill again and this made me unreliable (in his eyes). He made it very clear to me that I was an excellent lawyer who, fully fit, would have been in his first team every day of the week, but that my illness made it very difficult for him to manage the rest of the team and the workload. I understood his rationale perfectly and we managed to part company on very good terms, but I was totally shattered by the decision. And whilst I could have fought for my job on grounds of disability I didn't want to. I was exhausted by my illness and having been a team manager myself in previous roles I understood my line manager's predicament and may well have done the same thing in his position.

I think that part of the reason I was so devastated about leaving work was because it dawned on me that I wouldn't be able to apply for other jobs with my health the way it was at that time and my best chance of employment was to fight for my job with my then current employer. As I required frequent hospitalisations with IV fluids at the time and didn't really know why I couldn't honestly have looked a new prospective employer in the eye and apply for a decent legal position.

I've always been very ambitious and have excelled in a number of different roles since qualifying as a solicitor over sixteen years ago. Therefore coming to terms with the loss of opportunity with my career has been particularly tough. Indeed even after I'd accepted the proposal to terminate my employment it took me a long time to mentally let go of my career because throughout my repeat hospitalisations I'd always managed to contribute usefully, often working remotely from my hospital bed or from my office at home. Additionally, work had acted as a lifeline to normality during my illness because the law and working as a lawyer was a way of life for me throughout all my adult life.

There are, however, three definite positives that have come from leaving full time employment.

The first is that I've started working on a very part time basis for my husband's business as the in house legal counsel. This is not something I would have done had I not been forced to leave my last role – partly because I enjoyed my own independent career too much and partly because I earned a decent wage. But I've actually thoroughly enjoyed working in my husband's small business far more than I thought I would. I have renewed my practising certificate and managed to remain stimulated and keep my hand in. I work anything up to twenty hours a week and have complete autonomy and flexibility which is fantastic. I have also just joined a "virtual" law firm as a consultant and am starting to take on additional legal work outside of my husband's business. This is particularly exciting because I can work primarily from home and build my workload at my own pace and in line with what I feel I can take on at any given time given any childcare or health constraints.

The second, and by far the most important positive to come out of me giving up *"full time paid employment"*, is that I am now around far more for my girls. Previously when I'd been working full time I would often leave the house before the girls left for school and would come home after they were in bed. I would rely upon the school club and my nanny for wrap around childcare and would then just devote my weekends to my family. But as I became increasingly unwell I was no good to my family at the weekends as I started to spend more and more time in bed…….At least now when I'm well I can enjoy my children and my husband and I appreciate them so much more. I feel so blessed to have such an amazing husband and two such bright, articulate, intelligent and loving daughters with whom I can now spend far more quality time.

Then thirdly, leaving full time employment has enabled me to write this book which I've thoroughly enjoyed and which has been by far the best therapy I could have wished for…..

*But what about the financial impact of my ill-health?*

Well, the financial impact of my illness has been huge. Not only have I lost two jobs through ill health and missed out on a significant promotion I effectively had to kiss goodbye to a six figure earning capacity as a lawyer for a number of years. (Indeed I may never fulfil my earning potential as a lawyer again, depending on how my health pans out longer term.)

In addition to this loss of income I've had to bear increased childcare costs to care for the girls when I'm poorly; my private healthcare insurance premiums have increased over time and I've paid out many thousands over the years in insurance excesses, private consultations plus out of pocket expenses not covered by the insurance companies, such as take home medications and private ambulance costs. Indeed it was because of the very real and significant financial impact that my illness has had on me and my family that I started to pursue legal proceedings. But even if I had continued with the litigation and been successful, no amount of compensation would ever have been enough.

It is impossible to put a price on perfect health.

And the time I have lost with my family and the career opportunities I have missed are irreplaceable.

But I still remain hopeful and excited about my life and have faith that our healthcare system can and will improve. Clearly, listening to patients and their families more will, in my view, help enormously with any improvement.

# Hope for the future

My personal hope for the future is that my health will continue to improve and my need for IV hydration will decrease or even cease over time and that I can find a solution for my sexual dysfunction.

I had actually resigned myself to having to live a life without complete sexual satisfaction and it wasn't until I went out for dinner with a dear friend and discussed my frustration at not being able to reach such earth shattering climaxes following my second labour that she said:

*"You know you really shouldn't have to live with that. It's a quality of life thing. You're still so young and there may be a solution for you out there…… you need to look into it."*

And you know what? She's so right. Now that I'm no longer acutely ill and am moving into the phase of managing a long term chronic health condition I need to start looking at getting each and every area of my health as good as possible and that, of course, includes my sexual health.

Whilst the *"expert witness"* I consulted when I was considering legal action seemed to be convinced that my sexual dysfunction was almost certainly multifactorial in its aetiology I know for sure that it isn't. And in any event he wasn't a urogynaecologist. My sexual dysfunction started immediately after my second labour and it's as simple as that. Something happened to my body during that delivery which interfered with my ability to climax and I'm almost certain that it relates to the pudendal neuropathy I suffered, but rather frustratingly the SNS hasn't fixed this. It has helped a little with some recovery of sensation, but so far, not enough.

As for my sex life, I still find my husband very attractive and still totally enjoy sex, but I just can't come in the same way as I used to.

I have even done my fair share of soul searching and have wondered if there could be something missing in my marriage and what if I just had a random affair to see if I could come with a different partner? But I couldn't embark upon an affair as I still love and adore my husband. And even a vibrator (with a large amount of fanaticising about some other worldly sex symbol after a few glasses of vino) doesn't do the trick like it should...... ! So I'm 100% certain it's not in any way multifactorial – it's as a result of the birth trauma and that's that.

Instead I do think I will heed my friend's advice and next on my consultant shopping list is to visit an eminent urogynaecologist to find a fix for my predicament. And if there are any out there reading this book then please do get in touch, just in case I'm still not fixed by the time you're reading this sentence!!

As for the bigger picture of the healthcare system I do still have much hope.

Through my repeat hospitalisations I have come to realise that both the NHS and the private sector in the UK can be excellent at times – but they both certainly do have their problems.

From a patient perspective, aside from the waiting lists and lack of funding, my main issue with the NHS is that you're not guaranteed to see a consultant when you're admitted to hospital, nor is there any guarantee that a consultant will perform your operation/procedure. There is also too much emphasis on relieving the symptoms of a patient's illness/trauma/disease and not enough emphasis on persevering to find the cause. The net result of this is that patients can be re-admitted over and over before any meaningful diagnosis is given, and sadly too many will die before a cure for that diagnosis is prescribed. But on the upside for the NHS, there are always doctors on hand who can make decisions on your care if you take a turn for the worse whilst an inpatient.

In the private sector, the obvious positives are that you get seen quickly, by a consultant and can be treated quickly, by a consultant. In addition you get good food (when you're well enough to eat it), a private room with TV, Wi-Fi, pillows, duvet and an ensuite and can have visitors any time of the day or night.

But if you deteriorate when your consultant is not around, the RMO is virtually powerless to make decisions on your care. Because the private sector is entirely consultant led, any key decisions about your care have to be taken by your consultant. So, during in patient stays there have been times when I've been in obvious agony or have become feverish and visibly on a downward spiral and in need of diagnostic tests but the RMO hasn't been able to order the test without first speaking to my consultant. And whilst my consultant has usually been readily contactable there are times when he hasn't been easy to track down. This is because the private hospital I've spent most of my time in, like the majority of private hospitals in the UK, is located on a separate site and some miles from my consultant's NHS hospital. This means that as a private patient my consultant is often not present. Whereas in the NHS if your consultant isn't around, there will usually be a doctor within your consultant's team who is on hand to see you.

By far the most sensible arrangement I've encountered along my journey, however, is being a private patient on a private wing within an NHS hospital. This meant that my consultant was more often than not in the building and so if changes to my medication or decisions on diagnostic tests were required it happened more quickly than when I was at a private hospital on a separate site from my consultant's NHS hospital. Further, the private wing that I stayed at used the NHS hospital's facilities, including the operating theatres and diagnostic imaging tools, but as a private patient I paid a fee to the NHS for the privilege. This made perfect economic sense to me because a valuable funding stream was being introduced into the NHS by the private sector. (The only problem here was that the hospital being run by the NHS was so inefficient in its administration that it took a year for the bill to be issued to my insurers (whose contract with the hospital says that it has to be billed within six months of the date of care) and as a result the insurers were threatening to withhold payment for my stay ..... what a mess.)

We are also starting to see some other blurring of the NHS and private sector boundaries in the UK because NHS patients can "*choose and book*" to have procedures at private hospitals. Unfortunately, however, from speaking to some of the support staff such as phlebotomists and radiologists in my local private hospital the rise in number of NHS patients being admitted has increased their workload, but there has not been a like for like increase in resources and no doubt the cash strapped NHS is fleeced for the privilege. Further there is no adequate information system to enable the seamless and instantaneous sharing of information. So, when I get my urine sodium tested on the NHS by my GP (as my insurers won't pay for this as it's classed as routine monitoring) the results aren't available in my private hospital so I have to physically collect them and take them with me or email them or fax them over in advance of my arrival.

As for the NHS, the waiting times can be so long, particularly if you're a patient that requires input from a specialist team and you're not high priority. Had I not had private insurance I'd have waited forever for an inpatient assessment at an intestinal failure unit in connection with my euvolemia.

I don't have all the answers for making our healthcare system in the UK (both NHS and private) more "*fit for purpose*" but my strong belief is that by listening to patient experiences, healthcare leaders and politicians will get a better understanding of what needs to happen to improve the system. Those in retail and service industries constantly listen to their customers, asking for feedback and offering a generally excellent service. As a lawyer in private practice I was told that the client was always right and my time as an in house lawyer in the retail sector also taught me to start from the premise that the customer is always right. So healthcare needs to follow this lead and treat the patient with the same level of respect, reverence and customer service. If a bit more time is spent on this in the education of tomorrow's healthcare professionals then I absolutely believe that the patient experience will be immeasurably improved, as too will patient outcomes.

I now also passionately believe that there needs to be an integration of certain alternative medicine approaches with traditional western or allopathic medicine.

As for respect, reverence and customer service for the patient, a huge part of this is plain and simple *LISTENING*. The biggest improvement to healthcare in the UK, or indeed anywhere in the world, will (in my opinion) come if all medics listen more closely to patient histories and experiences.

I recently read an interview in a national daily newspaper with the father of a teenager who died from secondaries following bowel cancer. The teenager raised an amazing £5million for teenagers with cancer via a Facebook campaign in his dying weeks and was awarded a posthumous MBE from the Queen for his efforts.

But his father would rather have his son alive than be collecting his posthumous MBE.

Pointedly, his father believes he could have been saved had doctors listened to his concerns as a parent when his son's symptoms first manifested themselves. He took his son to the doctors and A+E around ten times before they screened his son for cancer; he was repeatedly told that his son was too young for bowel cancer and his concerns weren't listened to even though he was a survivor of bowel cancer himself and had Lynch Syndrome.

His father asked consultants and nurses to test his son, even showing them leaflets about Lynch Syndrome being hereditary and a cause of bowel cancer, but no one listened and when he was finally screened it was too late.

But why is it like this?

Why as patients or parents of patients do we have to fight to be heard?

I can't even remember the amount of times doctors have tried to tell me my pain is in my head or there is nothing wrong. And earlier this year my nine year old daughter was sent home twice by A+E doctors who told us that her swollen ankle was just a soft tissue injury.

She had five x-rays on two separate occasions at the same A+E department and when we questioned the A+E consultant we were told that there was nothing to worry about and that she should try and exercise the ankle and start walking on it as soon as she could. We were also told that as a *"failsafe"* a consultant radiologist would review the x-rays again in the morning and we would be contacted if there was anything to worry about. I even asked if I should call them just in case and I was reassured that I shouldn't and that I would be called if I needed to be.

A full week went by before we received a call from the same A+E department telling us we should go back because:

*"they thought the leg was fractured after all".*

When I returned to A+E and insisted on a further x-ray being taken so they could see if there was any deterioration due to the negligent advice we'd been given about trying to exercise the leg and the foot, even I could see from the x-rays that the area of the fracture had deteriorated. But more alarmingly I looked at the clinical notes on the screen and a full five days prior to the telephone call a consultant radiologist had reviewed the x-rays and remarked in the notes that the patient needed to be called back because he suspected:

*"a salter harris fracture of the distal tibia and a possible fracture of the fibula".*

I was furious.

Then to add insult to injury I was told that my daughter would have to wait a further ten days to see anyone in the fracture clinic but would be given a *"back slab"* plaster cast in the meantime. But I didn't let my anger show because I just needed them to get my little girl in a cast (as painlessly as possible) so that I could then take her home and organise some proper care.

So I promptly booked her in to see a leading paediatric orthopaedic surgeon privately as soon as possible and did some research at home and realised that if a salter harris fracture isn't put back into place within a short time frame lasting damage, including growth arrest can occur……. At the moment we know that there is a 30-40% chance that one of our daughter's legs could end up two centimetres shorter than the other and she'll have to be monitored for the next few years to rule this out. Early indications are unfortunately showing some growth arrest and that she will, in all likelihood require painful surgery after she's stopped growing.

The extent of the negligence in this instance is mind boggling, but again thankfully this is a physiological problem, and not pathological like the unfortunate young cancer suffer.

I have so many more stories from my close circle of family and friends who've also not been listened to by health care professionals and who instead have had to deteriorate significantly before they've received the right treatment, or worse, have lost loved ones, including a new born grandchild.

And the number of times I hear stories about patients having their sanity questioned when the medics can't work things out is quite simply appalling.

In terms of my health, whilst I'm left with a life-long health condition I'm incredibly fortunate because I have a fantastic consultant who I can contact at any time and who will organise my hospitalisation for IV fluids when required. If he's not contactable for whatever reason or there are no beds at my private hospital (which is relatively rare) then he has left a letter in my electronic notes at his NHS hospital under my hospital number directing the A+E doctors what to do if I present at A+E with low urine sodium.

Since putting these measures in place I've always received the urgent hydration required. So the system works well for me. Then, whenever I travel abroad I take my letters from the leading gastroenterologist who started me on the fluid replacement and restriction regime and who spells out in his letter, to any doctor attending to me, that if my urinary sodium is low I need IV hydration. I therefore don't have to live in fear of becoming poorly and *"falling between the gaps"*. And as I said earlier, I'm also hopeful that as time goes on my condition will stabilise further and I may eventually manage without IV hydration.

Additionally, because I now have such an extensive and complex history I find that doctors and other health care professionals tend to listen to me far more so than when I first fell ill. When I first became poorly and didn't have a diagnosis, doctors often questioned my mental health, rather than taking my physical ailments at face value. But because I can now speak with authority about my condition, and usually know more about it than those who are treating me for the first time, I find that when I ask for pain relief or antiemetics or present with nausea and acute pain I am listened to, taken seriously and treated with respect.

In fact aside from being physically and emotionally draining, my illness has also been both humbling and educational. I now passionately believe that if all healthcare professionals spent more time listening and taking each patient's concerns seriously then the quality of our healthcare system would be transformed overnight. Part of this vital listening process is also treating every single patient as a unique person.

The Secretary of State for Health quite rightly pointed out in a speech in early 2013 that:

*"every patient is a person: A person with a name. A person with a family. Not just a body harbouring a pathology; not a diagnostic puzzle; not a four hour target or an 18 week problem; not a cost pressure – and most certainly not a "bed-blocker""*.

So please, if you are a healthcare professional reading this book, keep listening to your patients (and their families) and remember the wise words of the Secretary of State above.

And if you're a patient or the parent or guardian of a sick child, you know your body and your child best, so please keep persevering until you find that listening ear.

## *Thank you for listening* ☺

www.ingramcontent.com/pod-product-compliance
Lightning Source LLC
Chambersburg PA
CBHW060845170526
45158CB00001B/238

www.ingramcontent.com/pod-product-compliance
Lightning Source LLC
Chambersburg PA
CBHW060845170526
45158CB00001B/238